THE COMPLETE IDIOT'S GUIDE® TO

Trigonometry

by Izolda Fotiyeva, PhD, and Dmitriy Fotiyev

ALPHA

A member of Penguin Group (USA) Inc.

The authors dedicate this book to their family and to all the known and unknown mathematicians who devoted their lives to explore the wonder of the world that mathematics is.

ALPHA BOOKS

Published by the Penguin Group

Penguin Group (USA) Inc., 375 Hudson Street, New York, New York 10014, USA • Penguin Group (Canada), 90 Eglinton Avenue East, Suite 700, Toronto, Ontario M4P 2Y3, Canada (a division of Pearson Penguin Canada Inc.) • Penguin Books Ltd., 80 Strand, London WC2R 0RL, England • Penguin Ireland, 25 St. Stephen's Green, Dublin 2, Ireland (a division of Penguin Books Ltd.) • Penguin Group (Australia), 250 Camberwell Road, Camberwell, Victoria 3124, Australia (a division of Pearson Australia Group Pty. Ltd.) • Penguin Books India Pvt. Ltd., 11 Community Centre, Panchsheel Park, New Delhi— 110 017, India • Penguin Group (NZ), 67 Apollo Drive, Rosedale, North Shore, Auckland 1311, New Zealand (a division of Pearson New Zealand Ltd.) • Penguin Books (South Africa) (Pty.) Ltd., 24 Sturdee Avenue, Rosebank, Johannesburg 2196, South Africa • Penguin Books Ltd., Registered Offices: 80 Strand, London WC2R 0RL, England

International Standard Book Number: 978-1-61564-144-4
Library of Congress Catalog Card Number: 2011938629

14   13   8   7   6   5   4   3

Interpretation of the printing code: The rightmost number of the first series of numbers is the year of the book's printing; the rightmost number of the second series of numbers is the number of the book's printing. For example, a printing code of 12-1 shows that the first printing occurred in 2012.

Printed in the United States of America

Publisher: *Marie Butler-Knight*

Associate Publisher: *Mike Sanders*

Executive Managing Editor: *Billy Fields*

Senior Acquisitions Editor: *Tom Stevens*

Development Editor: *Ginny Bess Munroe*

Senior Production Editor: *Kayla Dugger*

Copy Editor: *Jaime Julian Wagner*

Cover Designer: *Kurt Owens*

Book Designers: *William Thomas, Rebecca Batchelor*

Indexer: *Johnna VanHoose Dinse*

Layout: *Rebecca Batchelor, Ayanna Lacey*

Proofreader: *John Etchison*

Contents

Introduction

We've written this book so that you can master important trigonometry topics. All concepts in this book are carefully explained, important definitions and procedures are provided, and sample problems with step-by-step solutions appear in every chapter.

Trigonometry often seems imposing and daunting at first—in fact, one of the authors also found it this way when he was studying it years ago. However, a carefully paced approach with plenty of explanation and examples provides tremendous help in tackling trigonometry and becoming comfortable with all the topics. Such an approach helped one of the authors conquer his fear of trigonometry and we hope that, through this book, we will extend a helping hand to you, too. We've broken down all the concepts and sample problems in the book into bite-size chunks that are easy to grasp and understand.

The concepts pertaining to a specific trigonometry topic are thoroughly explained; all definitions, formulas, and procedures are clearly presented. Each concept and formula is reinforced with sample problems that enable you to practice what you learn.

Beginning with Chapter 4, each chapter contains five practice problems that cover the important concepts in the chapter. You learn best by doing. You will solidify your understanding of the chapter's material by trying to solve similar problems right after you study the material. Solve these problems, check your answers, and then compare your answers with the worked-out solutions in Appendix B. These practice problems will help you determine whether you have mastered the material completely or need further study.

Finally, each chapter wraps up with a summary of the key material and lists the main formulas. Review this list and you'll know the most important concepts and formulas of each chapter.

As you begin your studies, we would like to offer some suggestions for using this book and for being successful in achieving your goal of mastering trigonometry. Read each chapter with a pencil (or pen) in hand. Move through the sample problems with great care and take your time so you fully understand each and every step of the solutions. As you proceed through the book, don't get frustrated if you don't understand something after the first reading. Try again, and things will become clearer as you read and try to follow each step.

After you learn the material presented in a chapter, solve all of the practice problems. The more problems you solve, the better you will understand the topics.

How This Book Is Organized

This book is presented in six parts.

In **Part 1, The Basic Tools,** you will review important geometry topics involving angles, triangles, and circles that will help you better understand trigonometry concepts. We then provide a refresher of some major algebra topics, firm up your knowledge of functions and their inverses, and recall operations with complex numbers.

In **Part 2, Triangle Trigonometry,** you will get acquainted with the first perspective of trigonometry, trigonometric ratios, and learn how to solve a triangle. Two important trigonometry laws, the law of sines and the law of cosines, will also be introduced to help you deal with almost any problem related to triangles.

In **Part 3, Trigonometric Functions and the Unit Circle,** you will be introduced to the second perspective of trigonometry, using the unit circle; find techniques to define and evaluate trigonometric functions of any angle, from negative to positive infinity; and learn to cope with the degree and radian angle measures.

Part 4, Graphs of Trigonometric Functions, introduces you to the host of trig function "portraits"—their graphs. You will learn to deal with different graph alterations: shrinking, stretching, and shifting along both axes. You will also face the challenge of sketching the graphs of inverse functions and evaluating compositions of trig and inverse trig functions.

In **Part 5, Trigonometric Identities and Equations,** you will look with more depth at fundamental identities and use them to verify challenging identities. You will also apply many algebra techniques to solve trigonometric equations and learn to find general solutions for them.

In **Part 6, Polar Coordinates and Complex Numbers,** you will get a chance to face complex numbers once again and find how to present them geometrically and write them in polar form. We introduce formulas that will allow you to find powers and roots of complex numbers. You will wrap up your study by learning how to use a graphing calculator to solve many trigonometric problems.

Finally, this book wraps up with four appendixes. Appendix A serves as a glossary and provides the definitions of the key trigonometric terms and concepts. Appendix B features the answers to the practice problems found in the chapters. Appendix C reviews the core rules and formulas of trigonometry. Last but not least, Appendix D provides trigonometric tables to calculate the sine, cosine, tangent, cotangent, secant, and cosecant values of an angle given either in degrees or radians, along with information for conversion between degree values and numerical radian values.

Extras

Following are descriptions of the different sidebars you will find in this book.

DEFINITION

Trigonometry is full of terms, notions, and rules that require an explanation. In order to master trigonometry, you will need to have full command of them, which is what these sidebars will help you do.

WORD OF ADVICE

Frequently, you might wonder "How did they do this?" or "How did they obtain this expression from that one?" This type of sidebar will provide you with tips and shortcuts that will save you a lot of time and eliminate frustration.

DANGEROUS TURN

These sidebars will warn you about the most common and dangerous pitfalls that are capable of completely throwing you off track.

WORTH KNOWING

These little snippets will broaden your view of the topic at hand, while providing some useful knowledge, facts, and trivia. They will also help to look at trigonometry from a historical perspective and understand its connection to other mathematical disciplines.

Acknowledgments

We would like to take this opportunity and thank many people who made this book possible. Your help, comments, and criticism were extremely valuable to us.

Very special thanks are due to our literary agent, Marilyn Allen at Allen O'Shea Literary Agency, who brought this book project into our lives. For Izolda, this book is a second project with Ms. Allen.

We are very grateful to the staff of Alpha Books, especially to our acquisition editor, Tom Stevens, for his valuable advice, direction, and encouragement. We also want to thank our development editor, Ginny Munroe, and our production editor, Kayla Dugger, for their professionalism and commitment.

We are highly indebted to Boris Meyerson, who again agreed to technically review this book before we sent it to the editors. This is the second book in this series that Boris has helped with. Boris received a Master's degree in Electrical Engineering from Ural State Polytechnic University, Russia, and currently works as a manager in the information technology field. We are very grateful for his help and expertise.

Finally, we would like to thank our extended family for their continuous support, love, and belief in us.

Special Thanks to the Technical Reviewers

The Complete Idiot's Guide to Trigonometry was reviewed by experts who double-checked the accuracy of what you'll learn here, to help us ensure that this book gives you everything you need to know about trigonometry. Special thanks are extended to Boris Meyerson and Eric Stroh.

Trademarks

All terms mentioned in this book that are known to be or are suspected of being trademarks or service marks have been appropriately capitalized. Alpha Books and Penguin Group (USA) Inc. cannot attest to the accuracy of this information. Use of a term in this book should not be regarded as affecting the validity of any trademark or service mark.

The Basic Tools

You begin the study of trigonometry with a short review of some geometry and algebra topics that are useful in mastering the core trigonometry concepts.

First, you learn some general facts about triangles and circles and recall several important formulas and theorems.

Later in this part, you review the Cartesian coordinate system, basic facts about functions and their inverses, and operations with complex numbers so you are fully equipped to deal with different trigonometric topics.

What Is Trigonometry?

In This Chapter

- Looking back in time to find the roots of trigonometry
- What do trigonometry and calculus have in common?
- Using trigonometry for surveying and navigation
- Identifying trig functions using a triangle and a circle

Trigonometry is often a source of difficulty and fear for students. The grief they experience with this subject often overshadows the usefulness of trigonometry in the real world. In addition to not seeing the connections to real life or understanding where trigonometric concepts come from, many students perceive trigonometry as a totally separate subject in the mathematical world. As a result of not having any connection to previously acquired mathematical knowledge from geometry and algebra (which trigonometry is based on) or any connections to more advanced mathematical topics, students are baffled by the intricacies of trigonometry and question both its purpose and the need to study it.

However, it's not as bad as it might seem at first. Trigonometry's core concepts are relatively simple, and its topics are rooted in both algebra and geometry. With an easy-to-follow, step-by-step approach, even the most intimidating trigonometric problems can be broken down and solved via a handy set of mathematical tools.

In this chapter, we look at the roots of trigonometry and provide a brief insight into where trigonometric concepts came from, how they were discovered, and how their uses in the past relate to how they are currently applied. With this in mind, you should be able to see the value of studying trigonometry as an important branch of mathematics instead of thinking about it as a detached, separate subject.

We also show that the underlying trigonometric concepts are relatively simple and understanding them can make the study of trigonometry painless.

How It All Started

Trigonometry is an old branch of mathematics, and many people from different nations contributed to its knowledge base. The term *trigonometry* comes from the Greek word *trigonon*, meaning "triangle," and the Greek word *metria*, meaning "measurement." As the name suggests, early trigonometry developed from the study of right triangles by applying the relationships between the measures of sides and angles to the study of similar triangles.

The history of trigonometry dates back to the second millennium B.C.E., during the early ages of Egypt and Babylon. The ancient Egyptians and Babylonians had developed calculation methods that dealt with the ratios of the sides of similar triangles, all before trigonometry was formalized into a separate branch of mathematics. These two civilizations had no clear definitions of *trigonometric functions* but were able to use them unknowingly to their advantage.

DEFINITION

A **trigonometric function** is used to relate the angles of a triangle to the lengths of the sides of a triangle.

Trigonometry was then advanced by the Greek astronomer Hipparchus in the second century B.C.E. He came to be known as the father of trigonometry and also became the first person whose use of trigonometry was documented. Hipparchus compiled a trigonometry table that measured the length of the chord (which is a segment of a circle) whose ends intersect the various angles in a circle of a fixed radius r. He then used these trigonometric tables for computations related to astronomy.

Later, in the fourth and fifth century C.E., Indian mathematicians created a trigonometry system based on trigonometric functions instead of the chords of circles. The history of trigonometry also included Arabic astronomers in the ninth through the thirteenth centuries C.E. who compiled and improved on the studies of both the Greeks and Indians. Arabic mathematicians such as Al-Jayyani and Nasir al-Din al-Tusi revived the already-discovered trigonometric functions and added new ones to the mathematical base of knowledge.

In the thirteenth century, the Germans developed modern trigonometry by defining trigonometric functions as ratios of quantities rather than lengths of lines. In the eighteenth century, Leonhard Euler defined trigonometric functions in terms of complex numbers, which we will discuss at the end of this book. These late-stage discoveries helped tie trigonometry to advanced fields of mathematics such as calculus.

Even though there have been few developments in trigonometry since the eighteenth century, future mathematicians might still make some remarkable discoveries.

Where Ancients Used Trigonometry

Like many other branches of mathematics, trigonometry was created out of necessity. Egyptians used trigonometry for land surveying and the building of their magnificent pyramids. Babylonians used trigonometry for their astronomical calculations. One of the earlier uses of trigonometry was the *shadow table*.

The results of a shadow table were different for each civilization; this is because shadow lengths were dependent on the position of the sun in the geographical location where the observations were taking place. It is important to mention that these early civilizations were inadvertently developing a relationship between shadow lengths and the hour of the day as a trigonometric function. Thus, more than 3,000 years ago, humans used the notion of a function before even creating the term *function* or understanding what a function really was.

DEFINITION

A **shadow table** is used to record a particular hour of a day that coincides with a particular length of the shadow of a gnomon (a vertical stick). Shadow tables were used as early as 1500 B.C.E. by the Egyptians to calculate time. Later, the Indians and Greeks developed similar tables.

A **function** is a dependent relationship between quantities when every input has exactly one matching output.

Modern Uses of Trigonometry

Modern mathematics has extended the uses of trigonometric functions far beyond a simple study of triangles. This has made trigonometry indispensable in many other areas, such as the field of physics, which uses radiation, the propagation of light and sound, alternating current, and other periodic phenomena, which all involve trigonometric functions. The sine and cosine functions are useful in modeling cyclical trends, such as weather patterns or the seasonal variation of demand for certain items.

Trigonometry is also used for finding positions of objects in relation to other objects, which is applicable in navigation (particularly satellite systems such as GPS), astronomy, the naval and aviation industries, oceanography, land surveying, and cartography (creation of maps).

Digital imaging and three-dimensional graphics are another real-life application of trigonometry. Computer generation of complex imagery is made possible by the use of trigonometric functions that define the precise location and color of each of the points of the image being constructed. The image is then made detailed and accurate by a technique referred to as *triangulation*.

DEFINITION

A **triangulation** is the process of determining the location of a point by measuring angles to it from known points rather than measuring distances to the point directly.

The uses of spherical trigonometry (a sub-branch of general trigonometry that is studied in advanced math courses) include astronomy and long-range navigation. Nowadays, transcontinental aircraft attempt to fly a route according to the laws of trigonometry, as such a route provides the shortest path between any two points on the globe.

Trigonometry and Other Mathematics Disciplines

Calculus, linear algebra, and, in particular, statistics use trigonometry. The study of trigonometric functions leads to a deeper understanding of *periodic functions* and all functions in general.

DEFINITION

A **periodic function** is a function that repeats its values in regular intervals, which are called periods.

Often, applications of trigonometry in these subjects involve the so-called Fourier series, named after the eighteenth- and nineteenth-century French mathematician and physicist Joseph Fourier. The Fourier series have a surprisingly diverse array of applications in the study of vibration analysis, acoustics, seismology, optics, and electric power engineering.

Fourier's series are utilized in areas where the connection to mathematics is far from obvious. One popular example is digital compression—images, audio, and video data are compressed into much smaller file sizes that are easier and faster to send over the Internet.

Two Historical Perspectives of Trigonometry

There are two historical perspectives of trigonometry used to introduce the trigonometric functions. One perspective uses the right triangle, which has acute angles that are labeled as θ. The three sides of the triangle are labeled adjacent, opposite, and the hypotenuse.

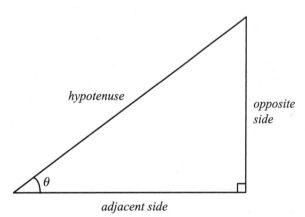

adjacent side

The right-triangle approach mimics the traditional approach to trigonometry used by ancient civilizations, as described earlier in this chapter. Using the lengths of the three sides, you can form six ratios that define the six trigonometric functions of the acute angle θ. But in this approach, you are limited to discussing acute angles, or angles measuring less than 90°. You learn more about this approach in Chapter 4.

The second perspective is based on the unit circle in the coordinate plane with center at (0,0) and radius unit of 1. The unit circle approach offers much greater flexibility because you can define trigonometric ratios for obtuse angles, which measure greater than 90°.

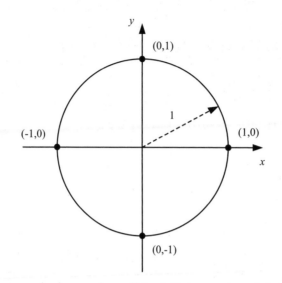

The unit circle approach also works well in defining and explaining trigonometric functions. Although it might seem less intuitive at first, many people like it and think that it illustrates trigonometric functions better than the triangle approach.

The only disadvantage of the approach is that while the sine and cosine functions appear very clearly on the unit circle, the other four trigonometric functions are challenging to spot at first glance. All six core trigonometric functions can easily be illustrated on a triangle, though.

As mentioned before, the unit circle definition does, however, have a large advantage in that it can define the trigonometric function for all positive and negative angles (even very large ones!), as opposed to just angles between 0° and 90°. We discuss the unit circle approach in more detail in Chapter 9.

The Least You Need to Know

- Trigonometry's history is rooted in the work of many individuals and cultures.
- The core relationships in trigonometry are ones between trigonometric functions and between the elements of triangles.
- There are two approaches for identifying trigonometric functions: the right-triangle approach and the unit circle approach.
- Trigonometry is an important part of many advanced fields of mathematics and has numerous applications in the real world.

Geometry Tools Needed to Study Trigonometry

In This Chapter

- What geometry cannot tell you about triangles and circles
- Finding a triangle's sides using the Pythagorean theorem
- Calculating the sides of special right triangles
- Calculating the measures of arcs in a circle

In this chapter, we look at some results and rules in geometry that set the stage for a successful study of trigonometry. Geometry concentrates heavily on two basic shapes: triangles and circles. For triangles, you find the lengths of a triangle's sides, its perimeter, and its area. For circles, you calculate the area and the circumference. However, geometry cannot always give us all the answers, and in those cases the powerful tools of trigonometry come into play.

Before delving into trigonometry's formulas and rules, let's recall some important facts and rules from geometry that you will need to explore this new discipline. We start our refresher with right triangles and with the most popular theorem on Earth—the Pythagorean theorem.

Right Triangles

Let's start our study of triangles with discussing the three main types of angles. These types are *acute angles*, *right angles*, and *obtuse angles*.

DEFINITION

An **acute angle** is an angle with a measure less than 90°.

A **right angle** is an angle with a measure of 90°.

An **obtuse angle** is an angle with a measure greater than 90°.

Recall from geometry that you need three known measures to describe a triangle—the lengths of two sides and the measure of an angle, the lengths of three sides, or three other measures. For a right triangle, you need only two additional measures, because you already know that one of the angles' measures is 90°.

The following four cases exhaust all possible ways to describe a right triangle:

1. The lengths of two legs

2. The lengths of one leg and the hypotenuse

3. The length of one leg and the measure of one acute angle

4. The length of the hypotenuse and the measure of one acute angle

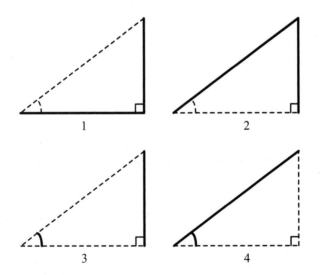

Suppose you need to calculate the lengths of all three sides. For case 1 and 2, you need only geometry, as these can be solved via the Pythagorean theorem. For cases 3 and 4, you need trigonometry since the relations of angle measures and side lengths involve trigonometric functions. We return to these cases in Chapter 4.

Let's consider another example. The measures of the sides of a triangle are 10, 12, and 14. What are the measures of its angles? It turns out that you cannot find the angles' measures using geometry. Sure, you can draw this triangle using the given sides' lengths and then measure the angles with a protractor. But there are no formulas or rules in geometry that allow you to calculate these angles. You need trigonometry to solve this problem. You also learn how to do this in Chapter 4.

The Pythagorean Theorem

Now let's look at the famous geometry tool, the *Pythagorean theorem*, which helps us to solve cases 1 and 2.

DEFINITION

The **Pythagorean theorem** states that if a and b are the lengths of the legs of a right triangle and c is the length of its hypotenuse, then $a^2 + b^2 = c^2$.

Let's illustrate how cases 1 and 2 can be solved with the help of the Pythagorean theorem.

Sample Problem 1

Two legs of a right triangle measure 5 and 12 inches. Find the length of its hypotenuse.

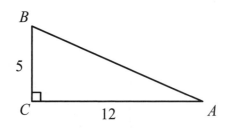

Step 1: Because you know the lengths of two legs, you can substitute their values into the left side of the theorem's equation:

$$a^2 + b^2 = c^2$$

$$(5)^2 + (12)^2 = c^2$$

Step 2: Calculate the left side of the equation:

$25 + 144 = c^2$

$169 = c^2$

The squared value of the hypotenuse length is 169.

Step 3: To find the length of the hypotenuse, extract the square root of 169:

$c = \sqrt{169} = 13$

Solution: The length of the hypotenuse is 13 inches.

Sample Problem 2

The hypotenuse of a right triangle is 5 inches long, and one leg is 4 inches long. Find the measure of the other leg.

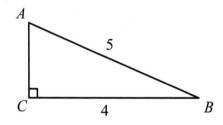

Step 1: Let's assume that the leg denoted by the letter a is 4 inches long. It is given that the hypotenuse c is 5 inches long. Therefore, you can substitute these values into the theorem equation:

$a^2 + b^2 = c^2$

$(4)^2 + b^2 = (5)^2$

Step 2: Calculate the left and right sides:

$16 + b^2 = 25$

Step 3: To find the unknown term b^2, subtract 16 from both sides:

$16 - 16 + b^2 = 25 - 16$

$b^2 = 9$

Step 4: To find the length of the leg b, extract the square root of 9:

$b = \sqrt{9} = 3$

Solution: The length of the leg b is 3 inches.

Special Right Triangles

In this section, we look at two special kinds of right triangles. For these special triangles, if you know the length of any one side of the triangle, you can find the length of the other two sides.

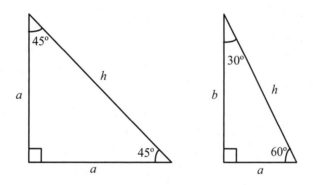

The triangle on the left is called a 45°-45°-90° triangle, or an *isosceles right triangle*. The triangle on the right is a 30°-60°-90° triangle. The Pythagorean theorem is used to prove the following relationships that exist in these two special right triangles.

DEFINITION

An **isosceles right triangle** is a triangle that has two equal sides and two equal angles, as well as a third angle that is a right angle. The only triangle that satisfies all of these conditions is a 45°-45°-90° triangle.

Theorem 1: If each acute angle of a right triangle measures 45°, the hypotenuse is $\sqrt{2}$ times as long as a leg.

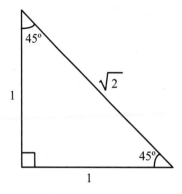

For example, if one leg in a 45°-45°-90° triangle is 10 units long, then the hypotenuse is $10\sqrt{2}$ units long.

Theorem 2: If acute angles of a right triangle measure 30° and 60°, then the hypotenuse is twice as long as the shortest leg and the longer leg is $\sqrt{3}$ times as long as the shortest leg.

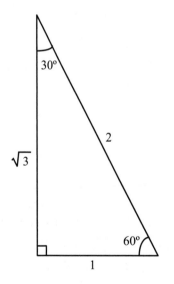

For example, if the shortest leg in a 30°-60°-90° triangle is 6 units long, then the hypotenuse is $2 \cdot 6 = 12$ units long, and the longest leg is $6\sqrt{3}$ units long.

Classifying Triangles

Any triangle has six elements: three sides and three angles. Let's agree to use capital letters A, B, and C to denote the measures of the angles. To denote the sides of the triangles, we use the small letters corresponding to the name of the angle opposite to this side. Let's also agree to use a symbol Δ to identify a triangle. Therefore, an expression ΔABC means "the triangle ABC."

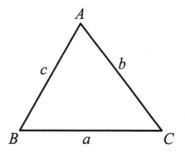

When given the lengths of three sides, you can construct a triangle that is determined by these sides; therefore, its angles are determined as well. However, you need to be careful in thinking this way, because not all arbitrary side lengths enable us to construct a triangle. Before constructing a triangle given its sides, you need to refer to the so-called *Triangle Inequality*.

DEFINITION

The **Triangle Inequality** states that the sum of the lengths of any two sides of a triangle is greater than the length of the third side.

Consider the following examples that make use of the Triangle Inequality. Let's state whether it is possible for a triangle to exist with sides of the given lengths.

Sample 1: The given sides are 4, 5, and 6 units long. You need to compare the sums of any two sides with the third side. That means you need to exhaust all possibilities:

$$4 + 5 = 9 > 6$$

$$4 + 6 = 10 > 5$$

$$5 + 6 = 11 > 4$$

Because the sums of any two sides are always greater than the third side, this triangle exists.

Sample 2: The given sides are 1, 2, and 3 units long. Let's try all possibilities:

$$2 + 3 = 5 > 1$$

$$1 + 3 = 4 > 2$$

$$1 + 2 = 3 = 3$$

The sum of these two sides (1 and 2) is not greater than the third side (3); therefore, this triangle does not exist.

A triangle can be classified as acute (having three acute angles), right (having one right angle), and obtuse (having one obtuse angle). As we discussed, the lengths of the sides of a triangle determine its angles. Given the lengths of the sides, can you tell whether the triangle is acute, right, or obtuse?

The Pythagorean theorem gives us a partial answer. From this theorem, you know that if the side lengths a, b, and c satisfy the relationship $a^2 + b^2 = c^2$, then the triangle is a right triangle. What if this relationship is not satisfied? Apparently, you will end up either with an acute or an obtuse triangle.

If $\angle C$ of $\triangle ABC$ is acute, then $c^2 < a^2 + b^2$

If $\angle C$ of $\triangle ABC$ is obtuse, then $c^2 > a^2 + b^2$

The following illustration summarizes these three possibilities:

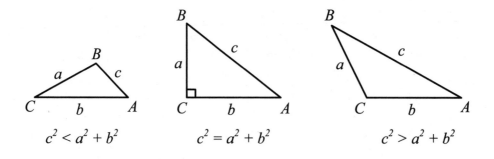

Using this information and given the sides' lengths, you can identify the type of triangle you are dealing with. For example, if the sides are 6, 7, and 8, by substituting those values, you get:

$$(8)^2 = 64 < (6)^2 + (7)^2 = 36 + 49 = 85$$

$c^2 < a^2 + b^2$, so the triangle is the acute triangle

Remember that, in a triangle, a hypotenuse is always bigger in length than either of the sides. Hence, we always know which number should be substituted for c since it's the largest one of the three.

If the sides are 6, 8, and 10, then

$$(10)^2 = 100 = (6)^2 + (8)^2 = 36 + 64 = 100$$

$c^2 = a^2 + b^2$, so the triangle is the right triangle

If the sides are 5, 12, and 14, then

$$(14)^2 = 196 > (5)^2 + (12)^2 = 25 + 144 = 169$$

$c^2 > a^2 + b^2$, so the triangle is the obtuse triangle

More on Triangles

In Part 2, we apply trigonometric functions to find a triangle's area, an angle's measure, and a side's length. Let's recall the formula for the area of a triangle.

Area of Triangles

The area of a triangle is equal to one half the product of the length of a base and the length of a corresponding altitude:

$$A = \frac{1}{2}bh$$

Recall that any side of a triangle can be taken as a base. There are three ways of expressing the area of the triangle, as shown in the following equation.

$$\text{Area of } \triangle ABC = \frac{1}{2}b_1h_1 = \frac{1}{2}b_2h_2 = \frac{1}{2}b_3h_3$$

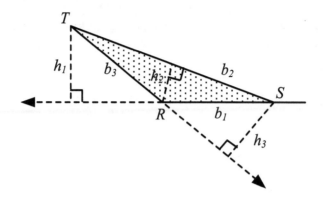

WORTH KNOWING

There are several formulas for calculating the area of a triangle, including ones that make use of coordinates or integrals. In this book, we learn about area formulas that use trigonometry and Hero's formula for the area of a triangle.

Conditions for Congruence

When any three elements of a triangle (out of six) are given, the triangle is uniquely determined. This fact is a consequence of the conditions for *congruence* that you learned in geometry. The table below summarizes four conditions for congruence.

DEFINITION

Congruence means that two triangles are the same shape and proportions.

Condition	Requirements
ASA Condition (Angle-Side-Angle)	If two angles and the included side of one triangle are congruent to the corresponding parts of another triangle, the triangles are congruent.
AAS Condition (Angle-Angle-Side)	If two angles and a nonincluded side of one triangle are congruent to the corresponding parts of another triangle, the triangles are congruent.

Condition	Requirements
SAS Condition (Side-Angle-Side)	If two sides and the included angle of one triangle are congruent to the corresponding parts of another triangle, the triangles are congruent.
SSS Condition (Side-Side-Side)	If three sides of one triangle are congruent to the corresponding parts of another triangle, the triangles are congruent.

We refer to these conditions later in Chapter 4 when we discuss how to find the elements of a triangle using trigonometric functions.

Circles, Arcs, and Chords

Another basic figure studied in geometry is a circle. Let's review some elements of a circle and formulas for the area and circumference of a circle. You'll need this information when solving problems using trigonometry. Let's first talk about the elements of a circle.

Chords and Arcs

Let's analyze the circle below and take note of its key elements. $\overline{OA}$ is a *radius*. $\overline{RS}$ is a *chord*. $\overline{BC}$ is a *diameter*.

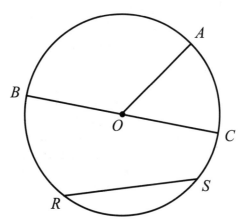

DEFINITION

A **radius** of a circle is a segment that joins the center of a circle to a point on the circle.

A **chord** is a straight line segment joining two points that lie on the circumference of a circle. A chord whose endpoints lie opposite to each other on the circle and whose center passes through the circle's center is referred to as the diameter.

A **diameter** is a chord that contains the center.

Next, let's discuss central angles and arcs of a circle.

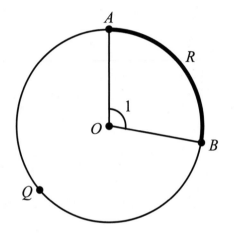

∠1 is the *central angle* of the circle *O*. The part of circle *O* shown in bold is a minor arc between the points *A* and *B*, which is denoted as arc *AB*. The part of circle *O* shown unbolded is *major arc AQB*. To name a major arc, three letters must be used.

The measure of a *minor arc* is defined to be the measure of its central angle. If the measure of $\angle 1 = 40°$, you can write that the measure of $AB = 40°$ as well. The measure of a major arc is calculated by subtracting 40 from 360 (the number of degrees in a full circle): $360° − 40° = 320°$.

DEFINITION

A **central angle** of a circle is an angle with a vertex at the center of the circle.

A **minor arc** of a circle is the union of two points on the circle and all the points of the circle that lie on the interior of the central angle with sides that contain the two points. Minor arcs measure less than 180 degrees.

A **major arc** is an arc of a circle having measure greater than or equal to 180 degrees.

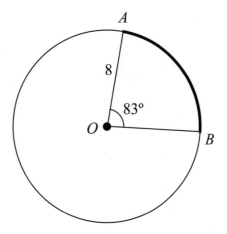

Let's consider another example with the circle. In a circle of radius 8, how long is the chord of an arc of 83°? It turns out again that geometry does not have the tools to solve this problem. You can draw the circle and the resulting triangle, but you cannot find the third side of this triangle using only geometry. This problem provides a glimpse of the types of problems that trigonometry will help you solve.

DANGEROUS TURN

If you are given the diameter D of a circle and need to calculate the area of a circle, don't forget to first divide the diameter by 2 to get the radius before using the area formula. Another option is to use the following formula for the area that includes the diameter:

$$A = \frac{\pi D^2}{4}$$

Area of Circles

To find the area of a circle, you need to know its radius. You can then use the following formula to calculate its area.

$$A = \pi r^2$$

The area of a circle is equal to the product of π and the square of the radius of the circle.

For example, if the radius is 7 inches, the area of the circle can be calculated by substituting this value into the area formula:

$$A = \pi r^2 = \pi(7)^2 = 49\pi$$

The Least You Need to Know

- To describe a triangle, you need to know three of its six elements.
- To describe a right triangle, you need to know only two of its elements.
- To decide whether a triangle exists, check its sides' lengths with the Triangle Inequality.
- There are three types of triangles: right, acute, and obtuse.
- The formula for the area of a triangle is $A = \frac{1}{2}bh$.
- The formula for the area of a circle is $A = \pi r^2$.

Algebra Tools Needed to Study Trigonometry

In This Chapter

- How trigonometry is connected to algebra
- Determining the signs of x and y in different quadrants
- Differences between functions and relationships
- Finding the inverse function
- Multiplying by the conjugate of a complex number

Because algebra and trigonometry deal with different areas of mathematics, many people think that trigonometry is completely divorced from algebra and that these two mathematics disciplines are independent of and separate from each other. But the truth is that you cannot understand trigonometry if you don't know algebra, making algebra a prerequisite to trigonometry.

Algebra deals with finding the value of unknown variables and understanding functions, whereas trigonometry explores aspects of triangles—sides and angles, and the relationship between them. Algebra seeks to solve equations composed of multiple terms and to find their roots, whereas trigonometry focuses more on sine, cosine, tangent, degrees, and radians.

It is helpful to think about trigonometry as an extension of algebra, at least to some degree, because we often treat trigonometric concepts in algebraic terms. For example, solving a trigonometric equation is similar to solving an algebraic equation. However, despite the similarities, there are also a number of differences between algebra and trigonometry. A thorough understanding of both is a prerequisite for more advanced mathematical topics such as calculus and differential equations.

In this chapter, we review some algebra concepts that will help in our exploration of trigonometric ideas.

Cartesian Coordinates

In algebra, you use the standard *Cartesian coordinate system* when drawing graphs of various functions. We discuss functions in the next section, but here we concentrate on the coordinate system itself. One of the common methods of solving systems of equations, the graphical method, involves graphing each equation on the same coordinate plane and finding the points of intersection of the graphs that correspond to solutions.

> **DEFINITION**
>
> The **Cartesian coordinate system** specifies each point on a plane by a pair of numerical coordinates, which are distances from the point to two fixed perpendicular lines, commonly referred to as the "x-axis" and the "y-axis."

The simplest coordinate system is a number line on which you represent numbers. Next, we represent pairs of numbers in a plane on which two of these lines are located.

In plane geometry, there are two axes at right angles that are usually called the *x*-axis and the *y*-axis. The position of any point in the plane can be given by its two coordinates (x,y). These coordinates give the point's distance in the x and y directions from the origin, which is the point of intersection of the two axes. The origin is labeled with the pair of numbers both at zero—$(0,0)$.

> **WORTH KNOWING**
>
> The Cartesian coordinate system was invented by the French mathematician and philosopher René Descartes in the seventeenth century. He linked geometry with algebra using the coordinate system and created a new mathematical discipline—analytic geometry.
>
> The coordinate system is called Cartesian because the idea was developed by René Descartes, who was also known as Cartesius. He is famous not only for his mathematical achievements but also for saying, "I think, therefore I am."

Writing Coordinates

You might have also heard the terms *abscissa* and *ordinate*—they are just fancy names for the x and y values. The following drawing illustrates the coordinate system with four labeled points. Note that, similar to the number line, you also can have negative values for x and y.

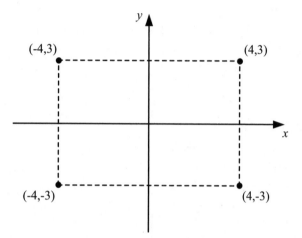

The origin $(0,0)$ is the starting point to determine distances when you need to graph other points. Two coordinates (x,y) are also called an ordered pair (a pair of numbers in a special order), because the order is important—the value for x always comes first, and the value for y comes second. Usually, the numbers are separated by a comma and enclosed by parentheses.

For example, to put a point on the coordinate system corresponding to ordered pair $(-5,3)$, count 5 units to the left from the origin on the x-axis because the x-coordinate has a negative value. For the y-coordinate, count 3 units up from the origin on the y-axis because its value is positive.

With the invention of the coordinate system, it became possible to visualize algebraic equations and plot their graphs. For example, a graph of a linear equation is a straight line, and a graph of a quadratic equation is a line called a parabola. We discuss graphs further in the next section of this chapter.

Four Quadrants

The *x*- and *y*-axes of the Cartesian coordinate system divide the plane into four parts, which are called *quadrants*. They are numbered in a counterclockwise direction using Roman numerals.

> **DEFINITION**
>
> **Quadrants** are the four regions into which the coordinate plane is divided by the *x*- and *y*-axes.

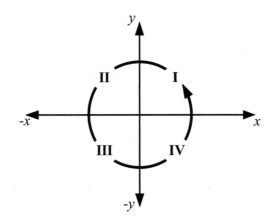

Note that the values for *x*- and *y*-coordinates are different for various quadrants in terms of their signs. For example, the values for the *x*-coordinate are positive in quadrants I and IV but negative in quadrants II and III. And the values for the *y*-coordinate are positive in quadrants I and II but negative in quadrants III and IV (see the following table).

This information is useful when considering the signs of trigonometric functions in different quadrants.

Quadrant	*x*	*y*	Example
I	Positive	Positive	(4,6)
II	Negative	Positive	(–2,10)
III	Negative	Negative	(–5,–8)
IV	Positive	Negative	(2,–9)

When you need to locate a spot in the real world, you have to use a three-dimensional coordinate system. In three dimensions, you have three axes at mutual right angles: x-, y-, and z-axes.

Coordinates can be continued into four and more dimensions, and mathematicians use these many-dimensional coordinate systems in their quest for new knowledge. Computers help them to visualize these many-dimensional worlds.

What Are Functions?

The concept of a function is one of the most important, fruitful, and indispensible tools in mathematics. The term *function* appeared for the first time in Gottfried Leibniz's mathematics manuscripts of 1673. The further development of the concept of a function is connected with Leonhard Euler's name, who was one of the greatest mathematicians of all time. Euler's work broadened the concept of a function and made it a core part of modern mathematics.

DEFINITION

A **function** is a correspondence or rule that assigns to every element in a set D exactly one element in a set R. The set D is called the domain of the function, and the set R is called the range.

In modern mathematics, functions continue to play a leading role; for example, function is a cornerstone of calculus. There is a branch of mathematics called complex analysis (also referred to as the theory of functions of a complex variable) that investigates functions of complex numbers. This part of mathematics is useful in many other branches of mathematics, including number theory and applied mathematics; it is also used in physics.

The Basics of Functions

There are different ways of looking at functions, and we'll consider a few here. But before we move forward and discuss some terminology, let's emphasize the fact that the function is not a graph, not a table of values for x and y, and not the equation. The function is a *relation* represented by these graphs, tables, and equations.

Let's illustrate the concept of a function with a real-life example. Note the names of the members of your family and your friends, and include their birthdays. The pairing of names and birthdays forms a relation. Note that, in this relation, as with functions, the pairs of names and birthdays are *ordered*. This example would imply that name always would be the first bit of information in the pair and birthday would be the second bit. The set of all the starting data (in our case, these are names) is called the domain and the set of all the ending data (in our case, these are birthdays) is called the range. The domain is what you start with; the range is what you end up with.

A function is not just any relation but only a *proper relation*. This means that, although all functions are relations because functions pair two sets of information, not all relations are functions. This means that functions are a subclassification (subset in mathematical terms) of relations.

DEFINITION

A **relation** is a relationship between two sets of information.

Entities are **ordered** if the order that they are presented in matters. Thus, if there are two entities, one comes first and the other comes second; this order should not be switched around.

A **proper relation** is a type of relation in which given an x, you get only and exactly one y.

Let's illustrate this with another example of family members and friends. As opposed to the previous case, let's flip things around and now assume that the domain (all values of x) is the set of everybody's birthdays. Let's imagine now that all family and friends have gathered for some event. There's a surprise-present-delivery guy waiting at the door, and all this fellow knows is that the surprise present is for the guest who has a certain birthday. Now you open the door and let him in. To whom does he give the present? What if nobody has this date as a birthday? What if there are four people in the room who have this date as their birthday? Do they all get a surprise present?

This means that the relation of "birthday indicates name" is not a proper relation as far as mathematics is concerned. It is not a function. Given the relationship $(x,y) =$ (name, birthday), if there are four people with the same birthday, there will be four different possibilities for $y =$ birthday. For a relation to be a function, there must be only and exactly one y that corresponds to a given x. Let's look at some examples that illustrate what we have just discussed.

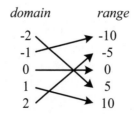

The preceding figure is definitely a function; you can trace from each x to each y. More importantly, there is only one arrow coming from each x.

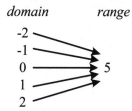

The second figure is different from the first example, but it is also a function. There is only one arrow coming from each x. The value for y is the same for each x, but it is only that one y. So this is a function, and this relation represents a horizontal line because the y value stays constant at 5 regardless of the x.

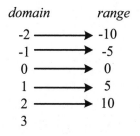

In this example, each x has its own y. Or does it? What about 3? It is a domain but it does not have any range element that corresponds to it. Thus, this one is definitely not a function. Moreover, it is not a relation either; the number 3 does not map to any other number.

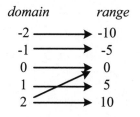

The relation expressed in the preceding figure is not a function because there are two arrows coming from the number 2. That means that the number 2 is connected with two different range elements. For one value of x, you have two values for y. Thus, it is not a function.

We write $y = f(x)$, where f is the symbol for the function itself.

There are unaccountably many different functions that are objects of study in such disciplines as real analysis and complex analysis, but we will focus on just the simpler ones in this book.

The Vertical Line Test

The graph of a function $y = f(x)$ consists of all points $(x, f(x))$ in a coordinate system. But not every equation in x and y is a function. Functions are a subset of the more general class of correspondence called relations. In mathematics, a relation is expressed by an equation in two variables—but not every equation represents a function.

For example, the relation described by the equation $y = 3x + 8$ is a function, because for every x you can find one and exactly one y. But the equation $y^2 = 5x - 7$ is not a function, because you cannot solve it for one unique y. There are two solutions—one positive and one negative:

$$y^2 = 5x - 7$$
$$y = \pm\sqrt{5x - 7}$$

For every value of x, you have two values of y; therefore, the equation does not represent a function.

The graph of the equation provides another way to determine whether an equation in x and y defines y as a function of x. This method is called the *vertical line test*.

DEFINITION

If no vertical line intersects a given graph in more than one point, then the graph is one of a function. This is called the **vertical line test.**

The graph on the left shows a function, because there is no vertical line that will cross the graph twice. The graph on the right does not represent a function, because any number of vertical lines will intersect this graph twice.

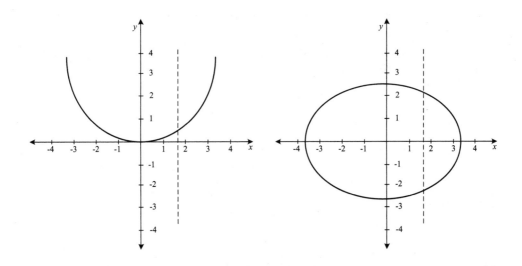

Inverse Functions and the Horizontal Line Test

Recall that you can measure temperature using two temperature scales: the Fahrenheit and the Celsius scales. The relationship between measurements in these two scales is as follows:

$$F° = \frac{9}{5}C° + 32$$

$$C° = \frac{5}{9}(F° - 32)$$

The first formula gives us a temperature in $F°$ as a function of $C°$, and the second one gives us a temperature in $C°$ as a function of $F°$. Because each formula undoes what the other one does, they are examples of *inverse functions*.

DEFINITION

Two functions f and g are called **inverse functions** if the following statement is true:

$g(f(x)) = x$ for all x in the domain of f

$f(g(x)) = x$ for all x in the domain of g

An inverse function is denoted by f^{-1} and read "f inverse."

The inverse of a function has all the same points as the original function, but the x's and y's are reversed. For example, if the original function has points {(2,3), (4,1), (6,–2)}, then the inverse function has points {(3,2), (1,4),(–2,6)}. In general, the graph of the inverse function f^{-1} can be obtained from the graph of f by changing every point (x,y) to the point (y,x). This means that the graph of f^{-1} is the reflection of the graph f in the line $y = x$.

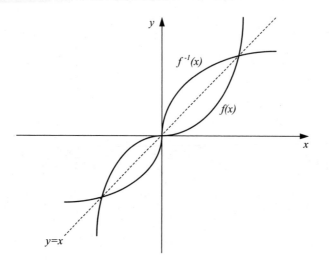

The *horizontal line test* is used to determine whether a function has an inverse that is also a function.

DEFINITION

If no horizontal line intersects a graph of a given function in more than one point, then this function has an inverse. This is called the **horizontal line test.**

Let's look at the example $f(x) = x^2$. The graph of this equation (on the left) is a u-shaped parabola that passes through the origin and opens upward. It is certainly a function, and it does pass the vertical line test. Note that the points (–1.3,1.69) and (1.3,1.69) are on this graph. Because a horizontal line can pass through the graph of $f(x) = x^2$ in more than one point, then $f(x)$ does not have an inverse that is also a function. On the other side, the graph on the right passes the horizontal line test, so it has an inverse function.

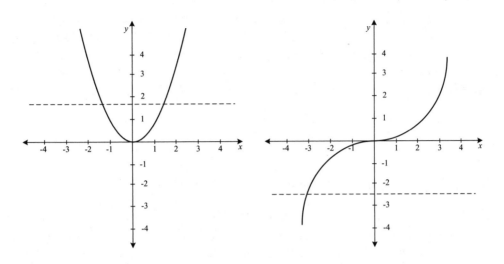

For some functions, it is easy to find their inverses using basic algebraic operations. As an example, let's find an inverse function for a function $f(x) = \dfrac{x-1}{2}$.

First replace the function notation with y: $y = \dfrac{x-1}{2}$. Then reverse the x's and y's: $x = \dfrac{y-1}{2}$. Now you need to solve for y:

$$2x = y - 1$$

$$y = 2x + 1$$

You find the inverse function $y = 2x + 1$ for the original function $y = \dfrac{x-1}{2}$.

Other more complicated functions such as e^x, $\sin\theta$, or $\ln x$ need more than a simple algebraic operation to find their inverse functions.

In this section we recalled some basic facts about functions. There are more things we need to know about functions for our study of trigonometry. We talk further about functions in Chapter 10 when we discuss periodic functions in connection to trigonometric functions, as well as transformation of functions (including shifting and stretching).

Complex Numbers

Our last stop on this algebra refresher is *complex numbers*. These numbers are important in modern mathematics and are used by mathematicians very often. They are also used in electric circuits and in the study of radio waves. So if it weren't for these numbers, we might not be able to talk on cell phones or listen to the radio!

> **DEFINITION**
>
> **Complex numbers** are numbers in the form $a + bi$, where a and b are real numbers, and i is the imaginary unit. a is the real part of such number and bi is the imaginary part of a complex number.

How They Came into Existence

Throughout the history of civilization, new kinds of numbers have been invented to fill deficiencies in the existing number system. In ancient times, there were only the numbers used for counting. Gradually, other rational numbers, such as fractions, were added. Then the Greeks discovered irrational numbers such as $\sqrt{2}$ or $\sqrt{7}$. Negative numbers were invented much later in Europe. All of these numbers comprise the real number system that includes zero, all positive and negative integers, and rational numbers and irrational numbers. One of the basic properties of all real numbers is that their squares are never negative.

In 50 C.E., Hero of Alexandria, a Greek mathematician and engineer, first encountered numbers with negative square roots when he studied the volume of a section of a pyramid. In order to calculate the volume, he had to determine $\sqrt{81-114}$. Because there was no way to determine the square root of a negative number at that time, Hero deemed this impossible and gave up. For a very long time afterward, no one tried to deal with these types of numbers.

In the sixteenth century, the concept of square roots of negative numbers was brought back. Formulas for solving third and fourth degree polynomial equations were discovered, and mathematicians realized that square roots of negative numbers were required. In 1545, Girolamo Cardano wrote a book titled *Ars Magna* where he made use of these strange numbers, but he greatly disliked them.

Many other mathematicians were slow to adopt the use of these numbers, including René Descartes, who didn't like theses strange numbers either. Nevertheless, in 1637, he developed the standard form for them, which is $a + bi$, where a is the real term and bi is the *imaginary number*. He also came up with the term *imaginary*, although he meant it to be derogatory. It took mathematicians more than two centuries to fully accept imaginary numbers and work out all the operations with them.

> **DEFINITION**
>
> An **imaginary number** is a number with square that is negative. Imaginary numbers have the form bi, where b is a nonzero real number and i is the imaginary unit, defined such that $i^2 = -1$.

Let's look at imaginary numbers more closely. When any real number is squared, the result is never negative. Imaginary numbers have the form bi, where b is a nonzero real number and i is the imaginary unit. Some examples of imaginary numbers are $2i$, $-5i$, and so on.

An imaginary number can be added to a real number a to form a complex number of the form $a + bi$, where a and bi are called, respectively, the *real part* and the *imaginary part* of the complex number. Therefore, pure imaginary numbers can be thought of as complex numbers where the real part equals zero.

With real numbers, we know that they all reside on the number line. Is there any place for complex numbers on the number line, or are they too strange for it? We come back to this question in Chapter 21.

Operations with Complex Numbers

You can add or multiply two complex numbers by treating i as if it were a variable and using algebraic distributive laws.

Sample 1: Add $5 + 4i$ and $9 - 2i$.

$$(5 + 4i) + (9 - 2i) = (5 + 9) + (4i - 2i) = 14 + 2i$$

Sample 2: Multiply $(2 - 3i)$ and $(5 + 4i)$.

$$(2 - 3i)(5 + 4i) = 10 + 8i - 15i - 12(-1) \rightarrow \text{ since } i^2 = -1$$

$$= 10 + 8i - 15i + 12 = 22 - 7i$$

The division of complex numbers is a little tricky. First we introduce *complex conjugate*s that help us perform the operation of division.

> **DEFINITION**
>
> The complex numbers $a + bi$ and $a - bi$ are called **complex conjugates.** Their sum is a real number, and their product is a nonnegative real number. The conjugate of the complex number $z = a + bi$ is denoted by $\bar{z} = a - bi$.

Let's illustrate with the following example that the sum of complex conjugates is a real number and that their product is a nonnegative real number.

Sample 3: Let's consider these complex conjugates: $3 + 5i$ and $3 - 5i$.

Their sum: $(3 + 5i) + (3 - 5i) = 3 + 5i + 3 - 5i = 6$

Their product: $(3 + 5i)(3 - 5i) = 9 - 15i + 15i - 25(-1) = 9 + 25 = 34$

Now you are fully equipped to divide two complex numbers.

WORD OF ADVICE

To divide complex numbers, multiply the numerator and the denominator by the conjugate of the denominator.

Sample 4: Divide $3 + 4i$ by $2 - 3i$.

$$\frac{3+4i}{2-3i} = \frac{(3+4i)(2+3i)}{(2-3i)(2+3i)} \rightarrow \text{note that the conjugate is } 2+3i$$

$$= \frac{6+9i+8i+12(-1)}{4+6i-6i-9(-1)} = \frac{6+17i-12}{4+9} = \frac{-6+17i}{13} = -\frac{6}{13}+\frac{17}{13}i$$

We learn more about operations with complex numbers in Chapter 21.

The Least You Need to Know

- Two coordinates (x,y) are called an ordered pair because the order is important; the value for x always comes first, and the value for y comes second.
- All functions are relations but not all relations are functions.
- To determine whether an equation in x and y defines y as a function of x, use the vertical line test.
- To determine whether a function has an inverse that is also a function, use the horizontal line test.
- Complex conjugates help in dividing complex numbers.

Triangle Trigonometry

This part introduces trigonometric functions using right triangles. You will become familiar with the six trigonometric ratios and understand their reciprocal relationships.

You learn how to write Pythagorean relationships using trigonometric functions and what it means to solve right triangles. You also learn how to find missing sides and angles.

Next, we discuss triangles, but this time we deal with oblique triangles. Because the Pythagorean theorem cannot help you solve oblique triangles, you learn the law of sines and the law of cosines, which enable you to accomplish that task.

While learning these laws, you will encounter an intriguing ambiguous case, where there is a possibility of having two different triangles with two given sides and an angle. You will learn how to determine what both of these triangles are.

The applications of trigonometry to navigation and surveying are discussed in this part as well.

Trigonometric Functions and Right Triangles

In This Chapter

- Using right triangles to identify trigonometric functions
- Introducing the tangent, the sine, and the cosine
- Relations between trigonometric functions
- Finding values for trigonometric ratios
- Finding angles of elevation and depression

There are two historical perspectives for introducing trigonometric functions. One of them uses right triangles and the other one uses the unit circle. This chapter focuses on the study of the first one. Trigonometric functions develop naturally from the study of right triangles, and this approach is useful for scientists and engineers to calculate the heights of extremely tall objects and distances that cannot be measured otherwise. This approach is also used to explain and predict many astronomical phenomena.

You usually measure distances using a ruler or a tape measure. However, you cannot use a ruler to measure a distance between a person standing on the beach and a ship in the sea. Such distances can often be found by using trigonometry rules. Remember, the word *trigonometry* comes from Greek words meaning "triangle measurements." In this chapter, we learn about the three special relationships that exist within any right triangle.

The Tangent Ratio

In the following figure, you can measure the distance BC on the beach. Can trigonometry help us to figure out what the measure of AC is?

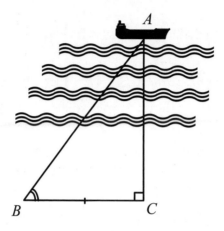

The answer is simple if $\triangle ABC$ is one of the special right triangles that we discussed in Chapter 2. Knowing the length of BC and $\angle B$, you can easily calculate the distance to the ship (AC). The next figure shows these triangles and the relations among the measures of their sides. We studied these relationships in Chapter 2.

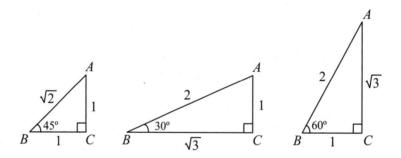

Note that for the 45°-45°-90° triangle, shown left-most on the figure, the ratio of its legs is $\dfrac{AC}{BC} = 1$. In other words, the ratio of the length of the leg opposite to the 45° angle to the length of the leg adjacent to the 45° angle is equal to 1.

For the middle triangle, the same ratio is $\dfrac{AC}{BC} = \dfrac{1}{\sqrt{3}} = \dfrac{\sqrt{3}}{3}$. In other words, the ratio of the length of the leg opposite to the 30° angle to the length of the leg adjacent to the 30° angle is equal to $\dfrac{\sqrt{3}}{3}$.

Finally, for the right-most triangle, the ratio is $\dfrac{AC}{BC} = \dfrac{\sqrt{3}}{1} = \sqrt{3}$. In other words, the ratio of the length of the leg opposite to the 60° angle to the length of the leg adjacent to the 60° angle is equal to $\sqrt{3}$.

Note that the second and the third triangles are both 30°-60°-90° triangles, but for the second triangle, $\angle B$ is 30°; for the third one, $\angle B$ is 60°.

How do you summarize the facts about these three ratios? If you denote all three acute angles as $\angle\alpha$, then you can state that the ratio of the length of the leg opposite to the $\angle\alpha$ to the length of the leg adjacent to the $\angle\alpha$ is equal to some number that is constant for each of three values of $\angle\alpha$. For 30°, this value is always $\dfrac{\sqrt{3}}{3}$, regardless of the lengths of the triangle's sides; for 45°, this constant is always 1; and for 60°, this constant is always $\sqrt{3}$. But how does this all help us to answer the initial question about the length of AC? To answer this, let's consider the following example.

Sample 1: For $\triangle ABC$ on the picture with the ship, let the length of BC be equal to 80 yards and $\angle B$ be equal to 30°. Then, according to what we just discussed, you can write the following:

$$\frac{AC}{BC} = \frac{1}{\sqrt{3}} = \frac{\sqrt{3}}{3}$$

Plug in the value for BC:

$$\frac{AC}{80} = \frac{\sqrt{3}}{3}$$

Cross-multiply:

$$3 \cdot AC = 80\sqrt{3} \approx 80 \cdot (1.73) \approx 138.56$$

Divide both sides by 3 to obtain the value for AC:

$$AC \approx 46 \text{ yards}$$

The distance to the ship is approximately 46 yards.

Similarly, you can find the distance to the ship (AC) for any other distance AB and for other angles (60° and 45°).

In summary, if you know the ratio of the length of the leg opposite to any acute angle to the length of the leg adjacent to this angle, you will be able to calculate the length of *AC* given the length of *CB*. That means you need to know these ratios for any acute angle.

WORD OF ADVICE

Using the trigonometric tables in Appendix D, locate the appropriate angle in the column labeled *Angle*. Read the value for tan in the column labeled *Tangent*. For example, for the 35° angle, the corresponding value of the tangent is approximately 0.7002. You can read more about the use of the tables in Appendix D.

Fortunately, these ratios for all acute angles have been calculated already and put into a special table called "Table of Trigonometric Ratios." You can find one in Appendix D. It is used extensively in the study of trigonometry. But before we move forward, we need to introduce an important term—the first truly trigonometric term in this book so far, by the way.

Did you notice that it was somewhat cumbersome to identify the ratio of two legs as "the ratio of the length of the leg opposite to an acute angle to the length of the leg adjacent to this angle"? Mathematics is a discipline that respects accuracy and conciseness. Certainly, mathematicians must have come up with some term to identify this ratio. This term was invented in the seventeenth century and became widely used due to the efforts of the great mathematician Leonhard Euler, whom we mentioned in Chapter 3.

The previously mentioned ratio between two legs of the triangle is called the *tangent* ratio and is denoted as tan.

DEFINITION

The **tangent** (tan) of an acute angle of a right triangle is the ratio of the length of the leg opposite to the acute angle to the length of the leg adjacent to the acute angle.

$$\tan \theta = \frac{\text{length of opposite side}}{\text{length of adjacent side}} = \frac{\text{opp.}}{\text{adj.}}$$

Explore the trigonometric table and notice that each acute angle corresponds to only one and exactly one value of tan. In other words, if the angles are different, then their tangents are also different. That yields special relations between the angle and its tan, the ones that we discussed in Chapter 3. Note that in this relation, as with functions, the pairs of angles and values for tan are ordered, which means one comes first and the other comes second. For example, you can set up this pairing so that either you choose values for acute angles, and then look up the corresponding values of tan, or you can choose tan values, and then look for the angles that yield these values.

Therefore, you can state that the tangent ratio is a function of an angle. This is the first trigonometric function that you have encountered; you will also learn about other trigonometric functions.

Remember that, for angle θ, the value of tan θ does not depend on the particular triangle that contains this angle; it depends only on the value for angle θ. Because tan is fundamentally a ratio, the lengths of the legs might be different and the length of the hypotenuse might be different; however, if the angle is the same, then the value for tan is also the same.

Let's now do some problems that use the tangent and see how it works.

Sample Problem 1

In the right triangle shown, find tan$A$.

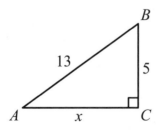

Step 1: According to the definition of tan, tan$A$ is defined as:

$$\tan A = \frac{\text{opp.}}{\text{adj.}} = \frac{5}{x}$$

So you first need to find x, which will then help you to find the tangent A.

Step 2: Find x by using the Pythagorean theorem:

$$c^2 = a^2 + b^2$$

$$(13)^2 = (5)^2 + x^2$$

$$169 = 25 + x^2$$

Subtract 25 from both sides to isolate x^2:

$$169 - 25 = 25 - 25 + x^2$$

$$144 = x^2$$

Extract the square root to find the value of x:

$$x = \sqrt{144} = 12$$

Thus, the length of the adjacent leg is 12.

Step 3: To find the value of $\tan A$, substitute 12 into the tangent ratio from Step 1:

$$\tan A = \frac{\text{opp.}}{\text{adj.}} = \frac{5}{x} = \frac{5}{12} \approx 0.4167$$

Solution: $\tan A = \dfrac{5}{12} \approx 0.4167$

Sample Problem 2

In the right triangle shown, find x correct to the nearest degree.

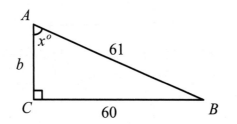

Step 1: You cannot use the tangent ratio because you don't know the length of the adjacent leg b (b denotes the leg opposite to $\angle B$, which in this case is between points A and C). Let's find it by using the Pythagorean theorem:

$$a^2 + b^2 = c^2$$

$$(60)^2 + b^2 = (61)^2$$

$$3{,}600 + b^2 = 3{,}721$$

$$b^2 = 3{,}721 - 3{,}600 = 121$$

$$b = \sqrt{121} = 11$$

The length of the leg b is 11.

WORD OF ADVICE

Remember that the sum of all angles in a triangle is 180°. This fact will be useful in solving many geometry and trigonometry problems.

Step 2: Knowing the length of both legs, you can find the tangent ratio:

$$\tan x = \frac{60}{11} \approx 5.4545$$

Step 3: Knowing the tangent x, you can use the trigonometric table in Appendix D and find the corresponding $\angle A$, whose measure is x degrees. The closest values for tangent in the table are 5.5301 and 5.3955, and correspond to angles 79.75° and 79.5°, respectively. The actual value of the angle thus lies between these two numbers. Because you are required to find the value of x to the nearest degree, you round the value for $\angle x$ up to 80°.

Solution: The value of x to the nearest degree is 80°.

Introducing the Sine and Cosine Ratios

The tangent ratio deals with the lengths of two legs. In this section, you learn other trigonometric ratios that relate to the hypotenuse as well. Let's look again at our right triangle and introduce the *sine* and *cosine* ratios:

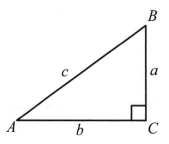

In the right triangle ABC, $\sin A = \dfrac{a}{c}$. This value depends only on the measure of the angle, and not on the lengths of the sides of the particular triangle we used.

 DEFINITION

The **sine** (sin) of an acute angle of a right triangle is the ratio of the length of the leg opposite the acute angle to the length of the hypotenuse:

$$\sin\theta = \frac{\text{length of opposite side}}{\text{length of hypotenuse}} = \frac{\text{opp.}}{\text{hyp.}}$$

The **cosine** (cos) of an acute angle of a right triangle is the ratio of the length of the leg adjacent to the acute angle to the length of the hypotenuse:

$$\cos\theta = \frac{\text{length of adjacent side}}{\text{length of hypotenuse}} = \frac{\text{adj.}}{\text{hyp.}}$$

Also, in the right triangle ABC, $\cos A = \dfrac{b}{c}$. Again, this value depends only on the measure of the angle, and not on the lengths of the sides of the particular triangle you used.

As with the tangent, the sine and the cosine ratios are functions of an angle. The values of $\sin\theta$ and $\cos\theta$ do not depend on the particular triangle that contains this angle; they depend only on the value for θ.

Cofunction Identity Between the Sine and Cosine

It's time to get to know sine and cosine more closely. In the figure, the following holds true: $\cos A = \dfrac{5}{9}$ and the value of $\sin B = \dfrac{5}{9}$.

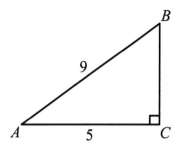

Is this a coincidence? Certainly, it is not. If $\angle A$ and $\angle B$ are acute angles of the same right triangle, then

$\sin A = \cos B$ and $\cos A = \sin B$

It doesn't matter what the lengths of the sides are.

Further observe that the sum of two acute angles in any right triangle is always 90° (because the third angle is 90°, and the sum of three angles is 180°), so you can express this fact with the following mathematical statement: $A° + B° = 90°$, which leads to the following conclusions:

$A° = 90° - B°$

or

$B° = 90° - A°$

Then you can rewrite the equalities from the previous paragraph such as:

$\sin A = \cos B = \cos(90° - A)$

$\cos A = \sin B = \sin(90° - A)$

These equations are called *cofunction identities*, and we discuss these further in Chapter 14 of this book.

DEFINITION

Cofunction identities for the sine and cosine can be written as

$\sin \theta = \cos\left(\dfrac{\pi}{2} - \theta\right)$ and $\cos \theta = \sin\left(\dfrac{\pi}{2} - \theta\right)$ where $\dfrac{\pi}{2}$ is equivalent to 90°.

Pythagorean Identity with the Sine and Cosine

Let's explore another relation step by step:

Step 1: According to the Pythagorean theorem, you write:

$$a^2 + b^2 = c^2$$

Divide both sides by c^2:

$$\frac{a^2 + b^2}{c^2} = \frac{c^2}{c^2}$$

$$\frac{a^2 + b^2}{c^2} = 1$$

Step 2: According to the definition of the sine and the cosine:

$$\sin A = \frac{a}{c}, \text{ then } \sin^2 A = \left(\frac{a}{c}\right)^2$$

$$\cos A = \frac{b}{c}, \text{ then } \cos^2 A = \left(\frac{b}{c}\right)^2$$

Step 3: Add the squares of the sine and the cosine from Step 2:

$$\sin^2 A + \cos^2 A = \left(\frac{a}{c}\right)^2 + \left(\frac{b}{c}\right)^2 = \frac{a^2}{c^2} + \frac{b^2}{c^2} = \frac{a^2 + b^2}{c^2}$$

Step 4: Compare the last lines in Steps 1 and 3. Because they are equal, you can write:

$$\sin^2 A + \cos^2 A = \frac{a^2 + b^2}{c^2} = 1$$

This expression is called the *Pythagorean identity*. We encounter more identities like this Chapter 14 of this book.

DEFINITION

The **Pythagorean identity** for the sine and cosine can be written as $\sin^2 \theta + \cos^2 \theta = 1$.

Values of Trig Functions for 0° and 90° Angles

So far we have identified the sine function only for an acute angle. Can we define sin90° and cos0°, for example? The following figure shows a series of triangles with the same hypotenuse but with an acute angle that increases.

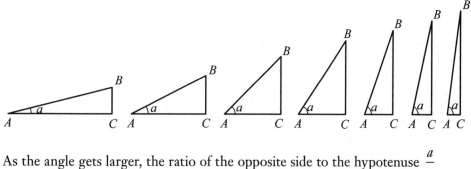

As the angle gets larger, the ratio of the opposite side to the hypotenuse $\dfrac{a}{c}$ approaches 1. Therefore, you can conclude that:

$$\sin 90° = 1$$

Similarly, you can explore the value for cos90°. As $\angle A$ gets closer and closer to 90°, the hypotenuse remains the same length, but the adjacent leg gets shorter and shorter. Thus, the ratio $\dfrac{b}{c}$ gets smaller and smaller. Therefore, you can conclude that cos90° = 0.

Let's explore now what happens when the $\angle A$ gets closer to 0° (the left side of the drawing). Apparently, the opposite leg becomes shorter and shorter and the ratio of the opposite side to the hypotenuse $\dfrac{a}{c}$ approaches zero. So you can assume that the sine of $\angle A$ approaches zero, and you can write that sin0° = 0.

At the same time, the adjacent leg becomes longer and longer, and the ratio $\dfrac{b}{c}$ gets larger and larger. Therefore, you can suggest that cos0° = 1.

Practical Applications of Trigonometric Ratios

The three trigonometric ratios are widely used by engineers and surveyors when they measure an *angle of elevation* or an *angle of depression*. For example, an angle of elevation is the acute angle through which a telescope must be elevated from the horizontal line to sight an object. The angle of depression is the acute angle through which a telescope must be depressed from the horizontal line to sight an object.

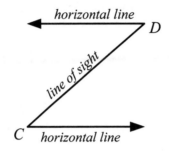

Note that $m\angle C = m\angle D$. $\angle D$ is the angle of depression for a person at D sighting C. $\angle C$ is the angle of elevation for a person at C sighting D.

DEFINITION

The **angle of elevation** of an object is the angle in a vertical plane between a horizontal line and the line of sight to the object. The **angle of depression** is equal to the angle of elevation.

Let's solve one problem and see how it works. The following problem is challenging; if you master this one, then other less challenging problems will be easy to solve.

Sample Problem 3

At a point A, a pilot flying toward an airport finds the angle of depression of the airport to be 22°. When the pilot flew 1,250 meters to point B, the angle of depression was 44°. If the pilot is flying at a constant altitude, what is the height of the plane above the ground?

Step 1: Let's denote the height of the plane above the ground as h. This is what the problem wants us to find. Make a drawing of the problem situation using the data from the problem:

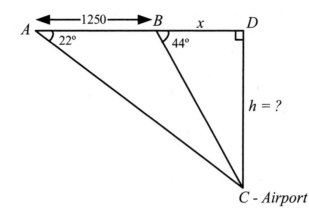

Step 2: Looking carefully, notice that there are two right triangles, $\triangle ADC$ and $\triangle BDC$. They share one leg that you denote as h. The other leg for both triangles is unknown because you don't know the distance BD. Let's denote this distance as x.

Note that you have two unknown variables, h and x. Recall from algebra that having two unknowns entails solving the system of two equations. But don't worry; we'll do it together, step by step.

Step 3: Let's first consider $\triangle BDC$. One of its legs is h, and the other leg is x. So you can write the tangent ratio for the 44° angle as:

$$\tan 44° = \frac{\text{opp.}}{\text{adj.}} = \frac{h}{x}, \text{ then}$$

$$x = \frac{h}{\tan 44°} = \frac{h}{0.9657}$$

$$x = \frac{h}{0.9657}$$

This is the first equation that you use in finding your answer.

WORD OF ADVICE

Because the expression $\tan \theta = \frac{b}{c}$ can be rewritten as $\frac{\tan \theta}{1} = \frac{b}{c}$ and cross-multiplied as $\tan \theta \cdot c = 1 \cdot b$, then $b = \tan \theta \cdot c$ and $c = \frac{b}{\tan \theta}$.

Step 4: Let's next consider $\triangle ADC$. One of its legs is still b, and the other leg is 1,250 + x. So you can write the tangent ratio for the 22° angle as:

$$\tan 22° = \frac{\text{opp.}}{\text{adj.}} = \frac{b}{1250 + x}, \text{ then}$$

$$1{,}250 + x = \frac{b}{\tan 22°} = \frac{b}{0.4040}$$

$$1{,}250 + x = \frac{b}{0.4040}$$

This is the second equation that you use.

Step 5: Write the system of two equations in two variables:

$$\begin{cases} x = \dfrac{b}{0.9657} \\ 1{,}250 + x = \dfrac{b}{0.4040} \end{cases}$$

You solve this system using the substitution method.

Step 6: Revise the second equation and substitute the value for x from the first equation:

$$0.4040(1{,}250 + x) = b \rightarrow \text{the revised second equation}$$

$$0.4040\left(1{,}250 + \frac{b}{0.9657}\right) = b \rightarrow \text{substituted expression for } x \text{ from the first}$$

equation

Thus, you obtain one equation in one variable, which should be fairly simple to solve using standard tools from algebra.

Step 7: Solve the equation. First distribute on the left side:

$$505 + \frac{0.4040b}{0.9657} = b$$

Simplify:

$$505 + 0.42b = b \text{ (you rounded the result of the decimal division to the}$$
hundredths)

Get the terms with the variable on the right side:

$$505 = h - 0.42h$$

Collect like terms:

$$505 = 0.58h$$

Divide both sides by 0.58:

$$h \approx 871$$

Solution: The height of the plane above the ground is 871 meters.

You should now be comfortable in understanding what the sine, cosine, and tangent functions are, as well as how to calculate them for various angles. As you saw in the preceding examples, these functions are very useful in calculating unknown angles and sides of a triangle that the standard tools of geometry were not helpful with.

Practice Problems

Problem 1: Without using the table, find the value of (tan30°)(tan60°).

Problem 2: In a right triangle, the hypotenuse is 17 units long. Find the length of a leg that is adjacent to an acute angle with a measure of 32°.

Problem 3: *ABCD* is an isosceles trapezoid. If *BC* is 16, *AD* is 30, and the measure of the ∠*C* is 110°, find *AB* correct to two digits.

Problem 4: A kite string is 250 meters long. Find the height of the kite if the string makes an angle of 40° with the ground. (Assume the kite string does not sag.)

Problem 5: From the top of a cliff 350 meters high, the angle of depression of a boat measures 26°. How far from the cliff is the boat?

The Least You Need to Know

- You can use right triangles to introduce the trigonometric functions.
- The tangent is defined as $\tan\theta = \dfrac{\text{opp.}}{\text{adj.}}$.
- The sine is defined as $\sin\theta = \dfrac{\text{opp.}}{\text{hyp.}}$.
- The cosine is defined as $\cos\theta = \dfrac{\text{adj.}}{\text{hyp.}}$.
- The Pythagorean identity for the sine and the cosine is $\sin^2\theta + \cos^2\theta = 1$.

Relations Among Trigonometric Ratios

In This Chapter

- Knowing reciprocal relationships
- Writing more Pythagorean identities
- Values of trigonometric functions for special angles
- Getting familiar with arcsine and other arcs

In this chapter, you continue working with right triangles and learn more about trigonometric functions and relations among them. We also discuss what it means to solve a triangle.

Six Trigonometric Relatives

You have studied three different trigonometric ratios: the sine, the cosine, and the tangent. These three ratios are closely related, and you have already seen that $\sin^2 \theta + \cos^2 \theta = 1$. It is time to introduce other members of the family of trigonometric ratios.

Given the $\triangle ABC$, you can identify them all:

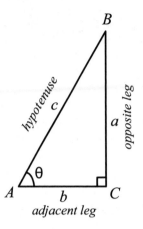

In any right triangle, we have a total of six different ratios of sides. We have already given names for three of them:

$$\sin\theta = \frac{\text{opp.}}{\text{hyp.}} = \frac{a}{c}, \quad \cos\theta = \frac{\text{adj.}}{\text{hyp.}} = \frac{b}{c}, \quad \tan\theta = \frac{\text{opp.}}{\text{adj.}} = \frac{a}{b}$$

The ratio between the adjacent leg and opposite leg is called the *cotangent* and is abbreviated as $\cot\theta$. Along with the previous three, these four trigonometric ratios are widely used. The two that are left are still used in the textbooks but are not so important mathematically. The ratio between the hypotenuse and the adjacent leg is called the *secant* and is abbreviated as $\sec\theta$. Finally, the ratio between the hypotenuse and the opposite leg is called the *cosecant* and is abbreviated as $\csc\theta$.

DEFINITION

Cotangent is a trigonometric function of an angle that calculates the ratio of the adjacent side divided by the opposite side of the angle, as follows:

$$\cot\theta = \frac{\text{adj.}}{\text{opp.}} = \frac{b}{a}$$

Secant is a trigonometric function of an angle that calculates the ratio of the hypotenuse divided by the adjacent side of the angle, as follows:

$$\sec\theta = \frac{\text{hyp.}}{\text{adj.}} = \frac{c}{b}$$

Cosecant is a trigonometric function of an angle that calculates the ratio of the hypotenuse divided by the opposite side of the angle, as follows:

$$\csc\theta = \frac{\text{hyp.}}{\text{opp.}} = \frac{c}{a}$$

Like the ratios for tangent, sine, and cosine discussed in the previous chapter, these ratios are dependent only on the size of the acute angle θ, and not on the lengths of the triangle's legs or hypotenuse.

Reciprocal Functions

From arithmetic, you might remember the term *reciprocal fractions*. To find the reciprocal of a fraction, we need to flip it over so that the numerator becomes the denominator and the denominator becomes the numerator. For example, the reciprocal of $\dfrac{3}{4}$ is $\dfrac{4}{3}$.

DEFINITION

Reciprocal fractions are fractions that represent $\dfrac{1}{x}$ of a given quantity of x. When multiplied by x, they give the product of 1.

Let's look at all six trigonometric functions, where a and b are the lengths of the triangle's legs and c is the length of the hypotenuse:

$$\sin\theta = \frac{a}{c}, \ \cos\theta = \frac{b}{c}, \ \tan\theta = \frac{a}{b}, \ \cot\theta = \frac{b}{a}, \ \sec\theta = \frac{c}{b}, \ \csc\theta = \frac{c}{a}$$

Have you noticed that the cotangent is the reciprocal of the tangent, the cosecant is the reciprocal of the sine, and the secant is the reciprocal of the cosine? You can summarize these facts by expressing them as:

$$\cot\theta = \frac{1}{\tan\theta}, \ \csc\theta = \frac{1}{\sin\theta}, \ \sec\theta = \frac{1}{\cos\theta}$$

These formulas are helpful when you want to prove the core *trigonometric identities* in Chapter 14.

DEFINITION

A **trigonometric identity** is an equation that is true for all values of the variable for which each side is defined.

Three Fundamental Identities

From the formulas for the reciprocal functions, you can derive many more identities. Fortunately, you don't have to remember them all. You need to remember only three fundamental ones, and all others can be easily derived from these three.

The first fundamental identity is the Pythagorean relationship between the sine and the cosine that we discussed in Chapter 4:

$$\sin^2\theta + \cos^2\theta = 1$$

For our next fundamental identity, consider the following:

$$\frac{\sin\theta}{\cos\theta} = \frac{\frac{a}{c}}{\frac{b}{c}} = \frac{a}{c} \div \frac{b}{c} = \frac{a}{c} \cdot \frac{c}{b} = \frac{a}{b} = \tan\theta$$

$$\tan\theta = \frac{\sin\theta}{\cos\theta}$$

For our third fundamental identity, recall first that the cotangent is the reciprocal of the tangent, so you can write:

$$\cot\theta = \frac{1}{\tan\theta}$$

Use the second fundamental identity:

$$\cot\theta = \frac{1}{\tan\theta} = \frac{1}{\frac{\sin\theta}{\cos\theta}} = \frac{\cos\theta}{\sin\theta}$$

WORTH KNOWING

Three fundamental trigonometric identities are:

$$\sin^2\theta + \cos^2\theta = 1, \tan\theta = \frac{\sin\theta}{\cos\theta}, \cot\theta = \frac{1}{\tan\theta} = \frac{\cos\theta}{\sin\theta}$$

Knowing these simple identities, you can derive many others that involve all six trigonometric functions.

Other Pythagorean Relationships

Notice that the first Pythagorean identity is also among the three fundamental trigonometric identities. Taking it as a base, we can perform some additional calculations in order to obtain other Pythagorean identities. By the way, they are not something new, they are just the rephrased first Pythagorean identity. So let's start with it:

$$\sin^2\theta + \cos^2\theta = 1$$

Divide each term by $\cos^2\theta$:

$$\frac{\sin^2\theta}{\cos^2\theta} + \frac{\cos^2\theta}{\cos^2\theta} = \frac{1}{\cos^2\theta}$$

Because $\dfrac{x^2}{y^2} = \left(\dfrac{x}{y}\right)^2$, you can rewrite:

$$\left(\frac{\sin\theta}{\cos\theta}\right)^2 + 1 = \frac{1}{\cos^2\theta}$$

The term in the parentheses is one of the fundamental identities, so you obtain:

$$\tan^2\theta + 1 = \frac{1}{\cos^2\theta}$$

Using the reciprocal function $\sec\theta = \dfrac{1}{\cos\theta}$ for the right term:

$$\tan^2\theta + 1 = \sec^2\theta$$

This is the second Pythagorean identity for trigonometric functions.

Now let's deal with the last Pythagorean identity. Start again with the first Pythagorean identity:

$$\sin^2\theta + \cos^2\theta = 1$$

This time divide each term by $\sin^2\theta$:

$$\frac{\sin^2\theta}{\sin^2\theta} + \frac{\cos^2\theta}{\sin^2\theta} = \frac{1}{\sin^2\theta}$$

You obtain:

$$1 + \cot^2 \theta = \frac{1}{\sin^2 \theta}$$

Using the reciprocal function $\csc \theta = \frac{1}{\sin \theta}$ for the right term:

$$1 + \cot^2 \theta = \csc^2 \theta$$

WORTH KNOWING

The three Pythagorean identities are:

$\sin^2 \theta + \cos^2 \theta = 1$, $1 + \tan^2 \theta = \sec^2 \theta$, $1 + \cot^2 \theta = \csc^2 \theta$

In Chapters 14 and 15, we verify many trigonometric identities that will require some tricks from algebra. In this section, let's verify some easy identities that will prepare us for more challenging problems.

Sample Problem 1

Prove that $\cot \theta \sin \theta = \cos \theta$.

Step 1: Because the three basic functions are the sine, the cosine, and the tangent, it is always useful to express all others through them. Recall that:

$$\cot \theta = \frac{\cos \theta}{\sin \theta}$$

Step 2: Plug in the value for the cotangent into original identity:

$$\frac{\cos \theta}{\sin \theta} \cdot \sin \theta = \frac{\cos \theta \cdot \cancel{\sin \theta}}{\cancel{\sin \theta}} = \cos \theta$$

$$\cos \theta = \cos \theta$$

Solution: You proved that $\cot \theta \sin \theta = \cos \theta$.

Let's do one more.

Sample Problem 2

Prove that $\dfrac{1-\sin^2\theta}{1-\cos^2\theta}=\dfrac{1}{\tan^2\theta}$.

Step 1: Let's work with the left side. From the first Pythagorean identity, you can obtain:

$$\sin^2\theta+\cos^2\theta=1$$
$$1-\sin^2\theta=\cos^2\theta$$

You will substitute the term on the right for the numerator in Step 3.

WORD OF ADVICE

An equation that is true for every number in the domain of the variable is an identity. Two examples of identities are $x^2-16=(x+4)(x-4)$ and $\dfrac{y}{4y^2}=\dfrac{1}{4y}$, $y\neq0$.

Step 2: Using again the first Pythagorean identity, you obtain:

$$\sin^2\theta+\cos^2\theta=1$$
$$1-\cos^2\theta=\sin^2\theta$$

You will substitute the term on the right for the denominator in Step 3.

Step 3: Plug in the right terms form Steps 1 and 2 for the numerator and the denominator:

$$\frac{1-\sin^2\theta}{1-\cos^2\theta}=\frac{\cos^2\theta}{\sin^2\theta}$$

Step 4: Notice that the term on the right side of the previous identity can be rewritten as:

$$\frac{\cos^2\theta}{\sin^2\theta}=\left(\frac{\cos\theta}{\sin\theta}\right)^2=\cot^2\theta$$

Step 5: Finally, recalling that the tangent is the reciprocal of the cotangent, you get:

$$\frac{\cos^2\theta}{\sin^2\theta}=\left(\frac{\cos\theta}{\sin\theta}\right)^2=\cot^2\theta=\frac{1}{\tan^2\theta}$$

Let's write all steps in one line, so it is well visible how you proved this identity:

$$\frac{1-\sin^2\theta}{1-\cos^2\theta} = \frac{\cos^2\theta}{\sin^2\theta} = \left(\frac{\cos\theta}{\sin\theta}\right)^2 = \cot^2\theta = \frac{1}{\tan^2\theta}$$

Solution: You proved that $\dfrac{1-\sin^2\theta}{1-\cos^2\theta} = \dfrac{1}{\tan^2\theta}$.

Determining Values of Trigonometric Functions

It is true that you can look up the values of trigonometric functions in the trigonometric tables found in Appendix D. However, you probably noticed that these values are approximate values, and you have to decide to what degree of accuracy you need to use the values of trigonometric functions.

For many problems you will encounter in the future, you will often find it is more convenient to use the exact values of trigonometric functions even if they contain radicals, or square root symbols. For example, in many calculations it is easier to use $\dfrac{\sqrt{3}}{2}$ for $\cos30°$ instead of referring to Appendix D and obtaining its approximate numerical value of 0.8660. It is practical to construct a table that will have trigonometric functions for the most often used angles. These angles are 0°, 30°, 45°, 60°, and 90°.

WORD OF ADVICE

To rationalize the denominator means to get rid of the radical in the fraction's denominator. To do so, multiply both the numerator and the denominator by the same radical expression and simplify. For example:

$$\frac{4}{\sqrt{2}} = \frac{4\cdot\sqrt{2}}{\sqrt{2}\cdot\sqrt{2}} = \frac{4\sqrt{2}}{2} = 2\sqrt{2}$$

You can construct this table using three pieces of information. The first piece of information is knowledge of the relations between the sides in special right triangles. Recall these facts:

	Opposite Leg	Adjacent Leg	Hypotenuse
Δ45°-45°-90°	1	1	$\sqrt{2}$
Δ30°-60°-90° for 30° angle	1	$\sqrt{3}$	2
Δ30°-60°-90° for 60° angle	$\sqrt{3}$	1	2

The second piece of information is the values for the sine and cosine of 0° and 90° angles. Let's recall them:

$$\sin 0° = 0 \qquad \cos 0° = 1$$

$$\sin 90° = 1 \qquad \cos 90° = 0$$

And the last piece of information you need for the construction of the table is the definitions of the trigonometric functions as discussed at the beginning of this chapter:

$$\sin\theta = \frac{\text{opp.}}{\text{hyp.}} \qquad \cos\theta = \frac{\text{adj.}}{\text{hyp.}} \qquad \tan\theta = \frac{\text{opp.}}{\text{adj.}}$$

$$\csc\theta = \frac{\text{hyp.}}{\text{opp.}} \qquad \sec\theta = \frac{\text{hyp.}}{\text{adj.}} \qquad \cot\theta = \frac{\text{adj.}}{\text{opp.}}$$

You are now fully equipped to construct the table using your knowledge of fractions and radicals. For example, to find the value of sin60°, plug in the values from the triangle sides relations table into the sine ratio:

$$\sin 60° = \frac{\text{opp.}}{\text{hyp.}} = \frac{\sqrt{3}}{2}$$

To find the value for the cosecant of 60° angle, plug in the corresponding values from the triangle sides relations table into the cosecant ratio:

$$\csc 60° = \frac{\text{hyp.}}{\text{opp.}} = \frac{2}{\sqrt{3}} = \frac{2 \cdot \sqrt{3}}{\sqrt{3} \cdot \sqrt{3}} = \frac{2\sqrt{3}}{3}$$

Note that for the cotangent, the cosecant, and the secant, you can use the reciprocal relations with the sine, the cosine, and the tangent. For example, let's find the same cosecant of 60° angle using reciprocal relations:

$$\csc 60° = \frac{1}{\sin 60°} = \frac{1}{\dfrac{\sqrt{3}}{2}} = \frac{2}{\sqrt{3}} = \frac{2 \cdot \sqrt{3}}{\sqrt{3} \cdot \sqrt{3}} = \frac{2\sqrt{3}}{3}$$

Let's set up a table with six columns and seven rows. The first column shows the type of function. Columns two through six show the values of functions for the five most commonly encountered angles:

Angle	0°	30°	45°	60°	90°
$\sin\theta$	0	$\dfrac{1}{2}$	$\dfrac{\sqrt{2}}{2}$	$\dfrac{\sqrt{3}}{2}$	1
$\cos\theta$	1	$\dfrac{\sqrt{3}}{2}$	$\dfrac{\sqrt{2}}{2}$	$\dfrac{1}{2}$	0
$\tan\theta$	0	$\dfrac{\sqrt{3}}{3}$	1	$\sqrt{3}$	undefined
$\cot\theta$	undefined	$\sqrt{3}$	1	$\dfrac{\sqrt{3}}{3}$	0
$\csc\theta$	undefined	2	$\sqrt{2}$	$\dfrac{2\sqrt{3}}{3}$	1
$\sec\theta$	1	$\dfrac{2\sqrt{3}}{3}$	$\sqrt{2}$	2	undefined

Some of the values for trigonometric functions are undefined. This always happens when you have to divide by zero. For example, to obtain the tangent of 90°, you have to divide by zero:

$$\tan 90° = \frac{\sin 90°}{\cos 90°} = \frac{1}{0} = \text{undefined}$$

WORTH KNOWING

The ancient Greek mathematician Hipparchus (190 B.C.E.–120 B.C.E.) created the first trigonometric tables. Today, scientific calculators have buttons for calculating the main trigonometric functions and their inverses. Most computer programming languages also provide function libraries that include the trigonometric functions.

After you put together the table, you can reflect on the question about the choice of our angles. Why these particular values? It is true that the chosen angles relate to the special triangles that you discussed many times already. Is there something else? Definitely, there is.

In Chapters 8 and 9, you deal with the second main approach in trigonometry of introducing the trigonometric functions using the unit circle. As with any circle, the unit circle can be divided evenly into 360°. Observe also that 360 can be evenly divided by 30, 45, 60, and 90, which are the measures of the angles in our table. This is handy when we discuss the multiples of the angles from the table and the values of their trigonometric functions in Part 3.

Solving Right Triangles

When some of the sides and angles of a triangle are known, the other elements can be found using trigonometry. This is called *solving a triangle*.

 DEFINITION

Solving a triangle is the process of finding the unknown elements of a triangle using the trigonometric relationships.

Let's learn how to find missing sides or missing angles in a triangle.

Finding the Sides

One of the most popular types of problems that students encounter in trigonometry textbooks is finding the shadows of objects. In solving these problems, you make use of the fact that the sun is very far away from the earth; therefore, you can consider that the rays of light from the sun reach a certain small area of the earth at the same angle. Let's do one of these problems.

Sample Problem 3

If you have a tree that is 20 feet tall, and the rays of the sun make a 24° angle with the ground, how long will be the shadow of the tree be?

Step 1: Make a drawing of the problem situation using the data from the problem:

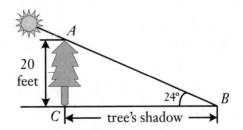

Step 2: On the drawing, AC is the tree and BC is the shadow. At the same time, AC is the opposite leg and BC is an adjacent leg of the $\triangle ABC$. Therefore, you have:

$$\tan 24° = \frac{AC}{BC} = \frac{20}{BC}$$

or

$$BC = \frac{20}{\tan 24°}$$

Step 3: Find the value of the tangent using either the calculator or the trigonometric tables in Appendix D and plug it into the equation from Step 2:

$$BC = \frac{20}{\tan 24°} = \frac{20}{0.4452} = 44.9$$

Solution: The length of the tree shadow is 44.9 feet.

Let's do one more problem on finding triangle sides.

Sample Problem 4

A bird sees breadcrumbs on the second-story windowsill from the top of a school flagpole at an angle of depression of 22°. The window is 18 feet above the ground and 50 feet from the flagpole. Find the height of the flagpole.

Step 1: Make a drawing of the problem situation using the data from the problem:

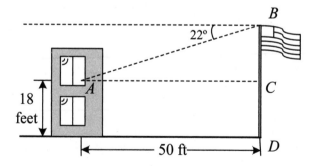

Step 2: The height of the pole BD is the sum of two parts BC and CD, one of which is unknown (BC). CD is 18 feet. You can express this statement using the data from the problem as:

Length of the pole $BD = BC + CD = BC + 18$

So to find the length of the pole, you need to find *BC*.

Step 3: Now let's deal with the angle of depression. Because it is 22°, the *complementary angle B* is 90° − 22° = 68°.

DEFINITION

Angles with a combined measure of 90° are called **complementary angles.**

Step 4: In $\triangle ABC$, you know the length of one side (*AC* = 50 feet) and the measure of $\angle B$, so you can find the length of *BC* as:

$$\tan 68° = \frac{AC}{BC} = \frac{50}{BC}$$

Then you can state that:

$$BC = \frac{50}{\tan 68°} = \frac{50}{2.4751} = 20.2$$

BC is 20.2 feet long.

Step 5: To find the height of the pole, plug in the value for *BC* from Step 4 into the expression for *BD* from Step 2:

$$BD = BC + CD = BC + 18 = 20.2 + 18 = 38.2$$

Solution: The height of the flagpole is 38.2 feet.

Finding the Angles

So far, when given the measures of angles and the lengths of some sides in right triangles, you found the lengths of unknown sides. But what if you need to find the angles? The first step is to write down the appropriate ratio between the sides of a triangle and get the value for the corresponding trigonometric function. Then you need to somehow get from this value to the measure of the angle itself.

WORD OF ADVICE

When calculators or trigonometric tables are used, only approximate values are available in most cases. Therefore, unless you are directed otherwise, it is standard practice to round the angle measures to the nearest tenth of a degree and give lengths accurate to three significant digits.

And that is what *inverse trigonometric functions* are all about—they provide the exact measure of the angles when the value of the trigonometric function is known. We discuss these in more detail Chapter 13.

There are two sets of notations for the inverse trigonometric functions that mean the same thing. On your calculator you should see, above the Sin, Cos, and Tan buttons, either notations Sin^{-1}, Cos^{-1}, and Tan^{-1}, or possibly ASIN, ACOS, and ATAN. You can read more about the usage of a graphing calculator in Chapter 22.

DEFINITION

Each trigonometric function has a corresponding **inverse trigonometric function.** The following notation is used:

$\sin^{-1}\theta$ or arcsin θ

$\cos^{-1}\theta$ or arccos θ

$\tan^{-1}\theta$ or arctan θ

and so forth

The first notation with the negative exponent lists the inverse sine, the inverse cosine, and the inverse tangent. The second set of notation, with an *A* before each name of the trigonometric function, stands for the arcsine, the arccosine, and arctangent, which are the full names for those three inverse functions. These are the same thing—and this is what you use to find angles based on the value of trigonometric functions.

We can state that $\sin^{-1}$ or arcsine means "the angle with a sine equal to ...." Let's do one problem and find the angle using the inverse trigonometric functions.

Sample Problem 5

A tall office building with a spire is 634 meters high. An observer at point *A*, 100 meters from the center of the building's base, sights the top of the spire. The angle of elevation is *A*. Find the measure of this angle to the nearest tenth of a degree.

Step 1: Make a drawing of the problem situation using the data from the problem.

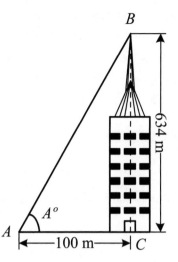

Step 2: Because you know the length of the opposite and adjacent sides of $\angle A$, let's use the tangent ratio:

$$\tan A = \frac{634}{100} = 6.34$$

Step 3: Find the measure of $\angle A$ using the tangent ratio:

$$\angle A = \tan^{-1} 6.34 \approx 81°$$

Solution: The angle of elevation is approximately 81°.

In this chapter, we have discussed the concept of reciprocal functions and introduced you to some of the key trigonometric identities. Coupled with the knowledge of the inverse functions, you now completed the foundation of your trigonometric "toolbox" that will help you in studying the upcoming chapters in this book as well as when working on many trigonometry problems.

Practice Problems

Problem 1: Prove that $\cos^2 \theta - \sin^2 \theta = 2\cos^2 \theta - 1$.

Problem 2: Prove that $\sin^2 \theta = \dfrac{1}{\cot^2 \theta + 1}$.

Problem 3: The hypotenuse of a right triangle is 12 and one of the acute angles is 37°. Find the other two sides.

Problem 4: A rectangle *ABCD* is 7 centimeters wide and 24 centimeters long. Find the measure of the acute angle between its diagonals.

Problem 5: A triangle has sides of lengths 16, 16, and 8. Find the measures of the angles of the triangle to the nearest tenth of a degree.

The Least You Need to Know

- Besides the sine, cosine, and tangent already introduced in Chapter 4, the other three trigonometric functions are as follows:

$$\cot \theta = \frac{\text{adj.}}{\text{opp.}} = \frac{b}{a}$$

$$\sec \theta = \frac{\text{hyp.}}{\text{adj.}} = \frac{c}{b}$$

$$\csc \theta = \frac{\text{hyp.}}{\text{opp.}} = \frac{c}{a}$$

- The three fundamental identities are as follows:

$$\sin^2 \theta + \cos^2 \theta = 1$$

$$\tan \theta = \frac{\sin \theta}{\cos \theta}$$

$$\cot \theta = \frac{1}{\tan \theta} = \frac{\cos \theta}{\sin \theta}$$

- The three Pythagorean identities are $\sin^2 \theta + \cos^2 \theta = 1$, $1 + \tan^2 \theta = \sec^2 \theta$, and $1 + \cot^2 \theta = \csc^2 \theta$.

- The inverse trigonometric functions help to find measures of unknown angles in right triangles.

The Law of Sines

In This Chapter

- Introducing oblique triangles
- Using the law of sines where geometry is helpless
- Dealing with an ambiguous case
- Deciding whether a triangle exists or not

We live our lives guided by laws, traditions, and culture. Some laws have been created by our Founding Fathers while some are very new. The same holds true with mathematics. New laws and theorems are created constantly in mathematics while others are covered by the dust of centuries that have passed. Nevertheless, they still work perfectly and their usefulness is unquestionable. In this chapter, we get acquainted with one of the workhorses of mathematics—the law of sines.

Oblique Triangles

So far, we have mostly dealt with right triangles, though we have also mentioned other triangles (especially in Chapter 2, where we reviewed some geometry topics). There are many other types of triangles that are classified depending on the specific features you want to emphasize.

For example, to stress that a triangle is definitely not a right one, you can call it an *oblique triangle*. If you want to define the oblique triangle further, you can say that it is either an acute or an obtuse one (refer to Chapter 2). Finally, you can reveal more details about the triangle and characterize it either as a *scalene triangle*, an *isosceles triangle*, or an *equilateral triangle*.

DEFINITION

Triangles that do not have a right angle are called **oblique triangles.** An **equilateral triangle** has three equal sides and three equal angles, always 60°. An **isosceles triangle** has two equal sides and two equal angles. A **scalene triangle** does not have equal sides or equal angles.

Note that right triangles also can be scalene (for example, the 30°-60°-90° triangle and many others), or isosceles (there is only one case—45°-45°-90°). But right triangles can never be obtuse or equilateral. For example, to be equilateral, the triangle must have all three angles be equal, but obviously all angles cannot be equal to 90°—and with the lesser degree measure, it stops being a right triangle.

Regardless of its specific features, each triangle still has the six elements that we previously discussed in Chapter 2. In this chapter, we concentrate on the oblique triangle and the trigonometric relations between its elements.

The Area of a Triangle

In Chapter 2, we mentioned that when the lengths of two sides and the included angle are known, the triangle is uniquely determined as a consequence of the SAS (Side-Angle-Side) condition of congruence. Let's express the area of the triangle using these measures.

Suppose that for $\triangle ABC$ you are given the lengths of its sides a and b and the measure of $\angle C$.

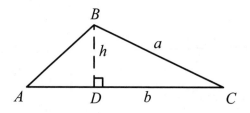

If the length of the altitude from B is h, then the area of the triangle can be expressed as:

$$A = \frac{1}{2}bh$$

Observe that the $\triangle BDC$ is the right triangle and that by right triangle trigonometry:

$$\sin C = \frac{h}{a} \text{ or } h = a\sin C$$

Substituting the last expression for b into the area of the triangle formula, you obtain:

$$A = \frac{1}{2} b \cdot (a \sin C) = \frac{1}{2} ab \sin C$$

Thus, you have the area of the triangle expressed through the sine of one of its angles. Similarly, if other pairs of sides and the included angles are known, the procedure can be repeated and two other area formulas can be obtained.

WORTH KNOWING

The area of $\triangle ABC$ is given by:

$$A = \frac{1}{2} ab \sin C = \frac{1}{2} bc \sin A = \frac{1}{2} ac \sin B$$

Notice that the pattern is always the same: half of the product of one side, the other side, and the sine of the included angle. This is a useful formula because, when these three elements are given, you don't have to look for the length of the altitude to find the triangle's area.

The Law of Sines

If each expression for the triangle area is divided by $\frac{1}{2} abc$, you obtain the *law of sines:*

$$\frac{\frac{1}{2} bc \sin A}{\frac{1}{2} abc} = \frac{\frac{1}{2} ac \sin B}{\frac{1}{2} abc} = \frac{\frac{1}{2} ab \sin C}{\frac{1}{2} abc}$$

$$\frac{\sin A}{a} = \frac{\sin B}{b} = \frac{\sin C}{c}$$

DEFINITION

In $\triangle ABC$, $\dfrac{\sin A}{a} = \dfrac{\sin B}{b} = \dfrac{\sin C}{c}$. This is called the **law of sines.**

The law of sines comes to the rescue when geometry becomes helpless. Suppose you have a triangle where you know two angles and one side, and you need to know the measures of other angles and the lengths of the remaining sides. Certainly, geometry can help you find the third angle (using the fact that the three angles of a triangle all sum up to 180°). But geometric methods don't enable you to compute the length of the other two sides. Trigonometry comes to help, specifically, with the law of sines.

Let's solve one problem and see the law of sines in action.

Sample Problem 1

A builder wants to measure the distances from points A and B to an inaccessible point C. From his direct measurements, he knows that AB measures 50 meters, $\angle A$ is 110°, and $\angle B$ is 20°. Find distances AC and BC.

Step 1: Make a drawing of the problem situation using the data from the problem:

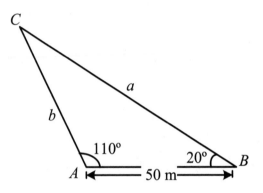

Step 2: Find the measure of $\angle C$:

$$\angle C = 180° - (110° + 20°) = 180° - 130° = 50°$$

The third $\angle C$ is 50°.

Step 3: Then, by the law of sines, you have:

$$\frac{\sin 110°}{a} = \frac{\sin 20°}{b} = \frac{\sin 50°}{50}$$

Let's first find the side a:

$$\frac{\sin 110°}{a} = \frac{\sin 50°}{50}$$

Cross-multiply:

$$50 \cdot \sin 110° = a \cdot \sin 50°$$

Divide both sides by $\sin 50°$, then:

$$a = \frac{50 \cdot \sin 110°}{\sin 50°} \approx \frac{50 \cdot 0.9397}{0.7660} \approx 61.3$$

Thus, BC is equal to 61.3 meters.

Step 4: Similarly, you can find AC. From Step 3:

$$\frac{\sin 20°}{b} = \frac{\sin 50°}{50}$$

Cross-multiply:

$$50 \cdot \sin 20° = b \cdot \sin 50°$$

Divide both sides by $\sin 50°$, then:

$$b = \frac{50 \cdot \sin 20°}{\sin 50°} \approx \frac{50 \cdot 0.342}{0.766} \approx 22.3$$

Thus, AC is equal to 22.3 meters.

Solution: The distance AC is 22.3 meters and the distance BC is 61.3 meters.

WORTH KNOWING

The law of sines was described in the thirteenth century by Nasīr al-Dīn al-Tūsī, a Persian mathematician and scientist. In his book, *On the Sector Figure*, he stated the law of sines for plane and spherical triangles and provided proofs for this law.

Notice that the law of sines is more general than the Pythagorean theorem. The Pythagorean theorem relates only to right triangles, but the law of sines is a theorem that can be applied to any triangle.

The Sine of an Obtuse Angle

Recall that we identified the trigonometric functions in a right triangle in terms of an acute angle. What about an obtuse angle? Can you identify its trigonometric functions? To discuss this, we begin with $\triangle ABC$ with the altitude denoted by h to side AB.

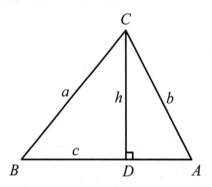

Using the sine ratio, you can express it in terms of the elements of the triangle. In fact, you can do it in two different ways using two right triangles—$\triangle ADC$ and $\triangle DBC$. From the first triangle, you get:

$$\frac{h}{b} = \sin A;\text{ therefore, } h = b\sin A$$

From $\triangle DBC$, you get:

$$\frac{h}{a} = \sin B;\text{ therefore, } h = a\sin B$$

Let's now consider an obtuse $\triangle ABC$ where $\angle B$ is an obtuse angle and CD is perpendicular to extended AB. Let's denote the altitude CD by h.

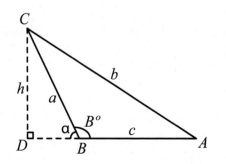

Like before, ΔADC is the right triangle, so you get:

$$\frac{h}{b} = \sin A; \text{ therefore, } h = b\sin A$$

By analogy with the previous example, you can expect that:

$$\frac{h}{a} = \sin B; \text{ therefore, } h = a\sin B$$

But there is no easy reference value for sine B because it is an obtuse angle. Those are commonly stated only for acute angles. To understand the situation, look at the ΔBDC. For this triangle, you can have:

$$\frac{h}{a} = \sin \alpha; \text{ therefore, } h = a\sin \alpha$$

If you compare two expressions, $h = a\sin B$ and $h = a\sin\alpha$, you can conclude that the sine of an obtuse angle is the sine of its *supplementary angle*.

DEFINITION

The sine of an obtuse angle is the sine of its supplement. Two angles are **supplementary angles** if their measure sums to 180°.

For example:

$$\sin 150° = \sin(180° - 150°) = \sin 30° = \frac{1}{2}$$

$$\sin 135° = \sin(180° - 135°) = \sin 45° = \frac{\sqrt{2}}{2}$$

What about other trigonometric functions of an obtuse angle? We discuss them in Chapter 9 when we present trigonometric functions using the second perspective with the unit circle.

The Ambiguous Case

The word *ambiguous* means open to or having several possible meanings or interpretations. But mathematics is a precise discipline with strict laws and rules where 2 + 2 is always 4.

Where is this ambiguity of a problem having multiple solutions coming from? Let's consider the following example to see how this ambiguity arises.

In $\triangle ABC$, $\angle B$ is 30°, side a is 4 centimeters, and side b is 3 centimeters.

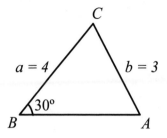

By the law of sines, you can find $\angle A$:

$$\frac{\sin A}{a} = \frac{\sin 30°}{b}$$

Cross-multiply:

$$b \cdot \sin A = a \cdot \sin 30°$$

$$\sin A = \frac{a \cdot \sin 30°}{b}$$

Substitute values for a, b, and sin30°:

$$\sin A = \frac{4 \cdot \frac{1}{2}}{3} = \frac{2}{3} \approx 0.667$$

Upon inspecting the trigonometric table found in Appendix D for the angle with a sine close to 0.667, you find that this angle is approximately equal to 42°. But from the previous section, you learned that supplementary angle 180° − 42° = 138° is supposed to have the same sine. That means that the problem has two solutions: one where $\angle A$ is an acute angle and another where $\angle A$ is an obtuse angle. The following figures illustrate these two possibilities.

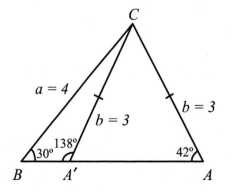

$\triangle ABC$ is the first solution with $\angle A$ being an acute angle, and $\triangle A'BC$ is the second solution where angle A' is an obtuse angle. Note that, in both triangles, side b equals 3 centimeters.

You are now ready to decide whether the triangle has only one solution, two solutions, or no solutions when two sides and the acute angle opposite one of them are given. There are three possibilities.

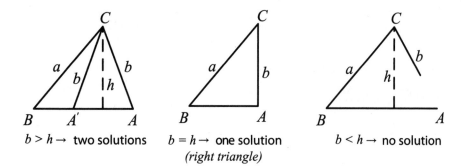

$b > h \rightarrow$ two solutions $b = h \rightarrow$ one solution $b < h \rightarrow$ no solution
(right triangle)

Let's show how it works with some examples.

Sample Problem 2

If a is 12 centimeters, b is 4 centimeters, and $\angle B$ is 60°, how many solutions are there?

Step 1: Make a drawing of the problem situation using the data from the problem:

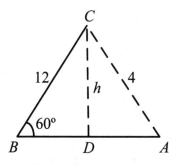

Step 2: In $\triangle DBC$, $\dfrac{h}{12} = \sin 60°$; therefore, $h = 12 \cdot \sin 60° = 12 \cdot \dfrac{\sqrt{3}}{3} = 4\sqrt{3} \approx 4 \cdot 1.73 \approx 6.93$.

Step 3: Because $b < h$ (4 < 6.93), the triangle does not exist. You can refer to the figure here and realize that if b is shorter than h, it would never be able to reach across between points C and A.

Solution: There is no solution and such a triangle does not exist.

DANGEROUS TURN

The sine of any angle cannot be greater than 1. If, while solving a triangle, you find that the sine of one of the angles is greater than 1, you need to check your work again.

Sample Problem 3

If a is 4 centimeters, b is 3.6 centimeters, and $\angle B$ is 45°, how many solutions are there?

Step 1: Make a drawing of the problem situation using the data from the problem:

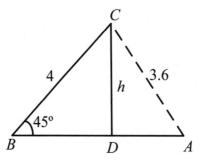

Step 2: In $\triangle BDC$, $\dfrac{h}{4} = \sin 45°$; therefore, $h = 4\sin 45° = 4 \cdot \dfrac{\sqrt{2}}{2} = 2\sqrt{2} \approx 2 \cdot 1.414 \approx 2.828$.

Step 3: Because $b > h$ (3.6 > 2.828), then the triangle has two solutions.

Solution: There are two solutions, since line b can be located on either side of where h is drawn on the illustration to this problem.

Solving Oblique Triangles Using the Law of Sines

In Chapter 2, we mentioned four conditions for congruence (SAS, SSS, ASA, and AAS). When these elements of a triangle are given, they uniquely determine a triangle—meaning that it has a unique shape and size. You also saw in the previous section that the case of SSA provokes the existence of two different triangles if the length of the side is greater than the height of the triangle. Let's illustrate all these possible cases with the following figure:

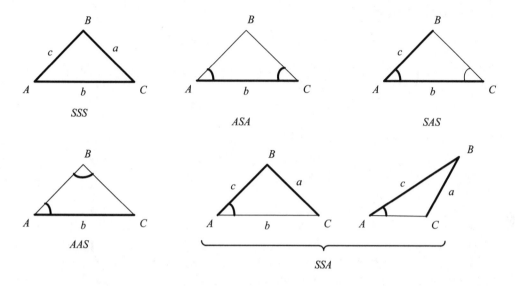

Let's organize what we already know about solving oblique triangles and summarize it in the following table. The first column lists what triangle elements are given, and the second column shows what geometry can help us to find. Finally, the third column shows where geometry falls short—and how the law of sines comes to the rescue to help find the missing elements.

Given	Geometry Can Help Find	Law of Sines Can Help Find
ASA	The third angle	The remaining sides
AAS	The third angle	The remaining sides
SSA	The included angle	An angle opposite a given side and the third side

WORTH KNOWING

The name *sine* comes from the Latin word *sinus,* a term related to a curve, fold, or hollow. Arab mathematicians in the eighth century transcribed the Sanskrit word *jya* (meaning a chord half) into *jaib,* which was written very similar to *jiba,* an Arabic word meaning a bay or a cove. Europeans mistakenly translated it into the equivalent Latin word for a bay or a cove—*sinus,* from which the modern term *sine* is derived.

Let's solve the last problem in this chapter using the law of sines to find the missing elements of a triangle.

Sample Problem 4

In a $\triangle ABC$, a is 18, c is 14, and $\angle A$ is 48°. Find the length of b.

Step 1: Make a drawing of the problem situation using the data from the problem:

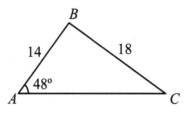

Step 2: To find the side b using the law of sines, you need to know the measure of $\angle B$. So far, you cannot identify it. But you find the measure of $\angle C$ using the law of sines. Why can the law of sines help you with $\angle C$ but not with $\angle B$? It is because the law of sines works with ratios of the sides and the sines of opposite angles. The problem provides you data to set up the following two ratios:

$$\frac{\sin A}{a} = \frac{\sin C}{c}$$

$$\frac{\sin 48°}{18} = \frac{\sin C}{14}$$

Cross-multiply to obtain:

$$18\sin C = 14\sin 48°$$

$$\sin C = \frac{14\sin 48°}{18} = \frac{{}^{7}\cancel{14}\cdot 0.7431}{{}^{9}\cancel{18}} = 0.5780$$

The sine of $\angle C$ is 0.5780.

Step 3: Knowing the sine of $\angle C$, you can find $\angle C$ using the inverse function:

$$\sin^{-1}0.5780 \approx 35°$$

The measure of $\angle C$ is 35°.

Step 4: Using geometry, you can find the measure of the third $\angle B$:

$$180° - (48° + 35°) = 180° - 83° = 97°$$

The measure of $\angle B$ is 97°.

Step 5: Because you know the opposite angle, you can now find the length of the side b using the law of sines. Note that $\angle B$ is an obtuse angle and its sine is equal to the sine of its supplementary angle. Let's find the supplementary angle:

$$180° - 97° = 83°$$

This is the angle you will use to calculate the length of b, because trigonometric tables provide the value of trigonometric functions for angles up to 90° only. If you want to use a graphing calculator, you can look up the sine value of 97° directly.

Step 6: Using the law of sines, you obtain:

$$\frac{\sin A}{a} = \frac{\sin B}{b}$$

$$\frac{\sin 48°}{18} = \frac{\sin 97°}{b}$$

Cross-multiply to obtain:

$$b\sin 48° = 18\sin 97° = 18\sin 83° \text{ (substituted with supplementary angle)}$$

$$b = \frac{18\sin 83°}{\sin 48°} = \frac{18 \cdot 0.9925}{0.7431} \approx 24$$

Solution: The length of the side b is 24.

You saw that the law of sines is useful for finding missing elements of a triangle, but it fails when you have the cases of SAS or SSS. It is clear that when, for example, three sides are given, no angles can be found using the law of sines since you cannot write any ratios between the sides and the opposite angles. But don't you worry, trigonometry has more resources to help; here is where the law of cosines can save the day. Our next chapter is about that law.

Practice Problems

Problem 1: The area of $\triangle RQP$ is 60. If p is 10, q is 20, and $\angle R$ is an obtuse angle, find the measure of $\angle R$.

Problem 2: Find $\sin 120°$, $\sin 125°$, and $\sin 145°$.

Problem 3: In $\triangle ABC$, $\angle A$ is 45°, a is 4, and $b = 2\sqrt{2}$. Find the number of solutions for the triangle.

Problem 4: In $\triangle ABC$, b is 7, $\angle A$ is 40°, and $\angle C$ is 28°. Find the length of a.

Problem 5: Use the law of sines to show that there is no $\triangle PQS$ with $\angle P$ equaling 30°, p equaling 3, and q equaling 8.

The Least You Need to Know

- When two sides and the included angles are given, the area of a triangle can be expressed by the following formulas:

$$A = \frac{1}{2}ab\sin C = \frac{1}{2}bc\sin A = \frac{1}{2}ac\sin B$$

- The law of sines provides special relationships between a triangle's sides and angles:

$$\frac{\sin A}{a} = \frac{\sin B}{b} = \frac{\sin C}{c}$$

- There is an ambiguous case that arises because an acute angle and a supplementary obtuse angle have the same sine.

- When you are given two sides of a triangle and an angle that is opposite to one of these sides, then 0, 1, or 2 triangles are possible.

The Law of Cosines

In This Chapter

- Defining the cosine of an obtuse angle
- Finding the triangle elements using the law of cosines
- Measuring the compass bearing
- Using the three sides to find the area of a triangle

In Chapter 6, we discussed some situations where the law of sines was unable to help you to find the missing elements of oblique triangles. This is where the law of cosines steps in to fix things up. The law of cosines is an old mathematical law; it is even older than the law of sines.

In this chapter, we introduce the law of cosines and discuss how it works and what kind of problems it helps to solve. We also apply both the law of sines and the law of cosines to tackle problems related to navigation and surveying. Finally, we learn another formula for the area of triangles.

The Law of Cosines

The law of cosines can be traced back to the third century B.C.E., when the great Greek mathematician Euclid of Alexandria included its geometric equivalent in his book *Elements* (the most famous book in geometry).

The development of modern trigonometry in the Middle Ages helped to formulate the law of cosines in its general form, and more progress was made in the fifteenth century when Jamshid al-Kashi, a mathematician from Samarqand, computed trigonometric tables to great accuracy and provided the first explicit statement of the law of cosines. From this time on, the law became suitable to use in *triangulation*, which we learn about later in this chapter. But only at the beginning of the nineteenth century did modern algebraic notation enable the law of cosines to be written in its current and familiar form.

> **DEFINITION**
>
> A technique that uses the law of sines and the law of cosines to compute the remaining elements of a triangle is called **triangulation.** This technique was commonly used for accurate, large-scale land surveying until the rise of global satellite-based positioning systems in the 1980s.

Proof of the Law of Cosines

The Pythagorean theorem is used to prove the law of cosines. Let ABC be a triangle with sides a, b, and c. Let $\angle C$ be an acute angle (you will understand later in this chapter why we make this specification for the angle).

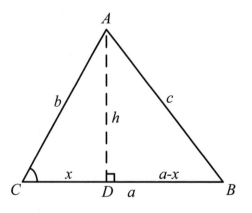

In order to prove the *law of cosines*, we want to show that $c^2 = a^2 + b^2 - 2ac\cos C$. Because trigonometric functions are defined in terms of right triangles, we resort to right triangles to prove the theorem. This is why we have the line AD, which is perpendicular to the side BC, and AD divides the original triangle, $\triangle ABC$, into two triangles, $\triangle DCA$ and $\triangle BDA$.

> **DEFINITION**
>
> The **law of cosines** states that in $\triangle ABC$, if $\angle C$ is an acute angle, then $c^2 = a^2 + b^2 - 2ac\cos C$.

Step 1: Denote CD as x, then BD is the side a minus segment x: $BD = a - x$.

Step 2: Let's first consider ΔDCA. In this triangle, because you know that $\dfrac{x}{b} = \cos C$, then it becomes clear that:

$$x = b\cos C$$

Also, according to the Pythagorean theorem:

$$b^2 = x^2 + h^2, \text{ then}$$
$$h^2 = b^2 - x^2$$

Step 3: In the right triangle ΔBDA:

$$c^2 = h^2 + (a - x)^2$$

Use the square of a binomial formula (which is $[a - x]^2 = a^2 - 2ax + x^2$) to expand and simplify:

$$c^2 = h^2 + a^2 - 2ax + x^2$$

Step 4: For h^2 in the previous equation, substitute the last expression from Step 2:

$$c^2 = h^2 + a^2 - 2ax + x^2 = \overbrace{b^2 - x^2}^{h^2} + a^2 - 2ax + x^2$$

Terms with x^2 cancel out each other:

$$c^2 = b^2 + a^2 - 2ax = a^2 + b^2 - 2ax$$

Step 5: Finally, substitute for x the expression $x = b\cos C$ from Step 2:

$$c^2 = b^2 + a^2 - 2ax = a^2 + b^2 - 2a\overbrace{b\cos C}^{x}$$

Thus, you have just proved the law of cosines.

In the same way, you can prove it for other sides of the triangle using the pattern:

(side opp. angle)² = (side adj. to angle)² + (side adj. to angle)² – 2(side adj. to angle)(side adj. to angle)cos(angle)

The Law of Cosines for an Obtuse Triangle

In the previous section, we emphasized that $\angle C$ is an acute angle. What if it is an obtuse angle? Will the law of cosines work for it as well?

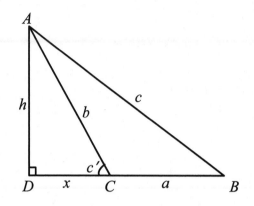

Let's show that if $\angle C$ is an obtuse angle, then:

$$c^2 = a^2 + b^2 + 2ab\cos C'$$

where C' is the measure of the supplement of obtuse angle C.

You can resort to the right triangle to prove this case and draw the perpendicular AD to the extended side BC. Thus, you again have two triangles, $\triangle CDA$ and $\triangle BCA$.

> **WORD OF ADVICE**
>
> The law of cosines looks like the Pythagorean theorem except for the last term:
>
> $c^2 = a^2 + b^2 - 2ab\cos C$
>
> Note that if $\angle C$ is a right angle, the last term disappears because the cosine of 90° is zero. Thus, the law of cosines is actually a generalization of the Pythagorean theorem.

Step 1: Denote CD as x, then BD is the side a plus segment x:

$$BD = a + x$$

Step 2: Let's first consider $\triangle CDA$. In this triangle, because $\dfrac{x}{b} = \cos C'$, then:

$$x = b\cos C'$$

Also, because $\dfrac{h}{b} = \sin C'$, then:

$$h = b\sin C'$$

Step 3: In the right triangle BDA:

$$c^2 = h^2 + (a + x)^2$$

Use the square of a binomial formula to expand and simplify:

$$c^2 = h^2 + a^2 + 2ax + x^2$$

Step 4: For h and x in the previous equation, substitute corresponding expressions from Step 2:

$$c^2 = h^2 + a^2 + 2ax + x^2 = (b\sin C')^2 + a^2 + 2a(b\cos C') + (b\cos C')^2$$

$$c^2 = b^2\sin^2 C' + a^2 + 2ab\cos C' + b^2\cos^2 C'$$

Step 5: Observe that the sum of the first and the last terms can be simplified using the fundamental identity $\sin^2\theta + \cos^2\theta = 1$ (refer to Chapter 4):

$$b^2\sin^2 C' + b^2\cos^2 C' = \text{take out the GCF} = b^2(\sin^2 C' + \cos^2 C') = b^2 \cdot 1 = b^2$$

Therefore, you can finally write:

$$c^2 = b^2 + a^2 + 2ab\cos C' = a^2 + b^2 + 2ab\cos C'$$

This proves the law of cosines for an obtuse triangle.

When you compare two formulas for an acute angle C and an obtuse angle C, you can suggest one interesting thing about the cosine of an obtuse angle:

$$c^2 = a^2 + b^2 - 2ab\cos C \qquad\qquad c^2 = a^2 + b^2 + 2ab\cos C'$$

Acute angle C \qquad\qquad\qquad Obtuse angle C

Because the left sides are the same and the first two terms on the right sides are the same, you can conclude that the third terms on the right sides should be equal as well. Therefore, you can state that:

$-2ab\cos C = +2ab\cos C'$

Because factors $2ab$ are the same, then $-\cos C = +\cos C'$, or $\cos C = -\cos C'$.

WORTH KNOWING

The cosine of an obtuse angle is the cosine of its supplement multiplied by -1.

We discuss the trigonometric functions of obtuse angles in Chapter 10.

Finding Angles with the Law of Cosines

The formula of the law of cosines can be changed to find the measures of a triangle's angles when the lengths of three sides are given. This formula is:

$$\cos C = \frac{a^2 + b^2 - c^2}{2ab}$$

Note two important things. First, this is not a new formula; this is the same law of cosines written in another way. Second, we provided this formula only for $\angle C$, but it can be used to find the other two angles by substituting appropriate values for the sides. The following is the general pattern for any of the triangle's three angles:

$$\cos(\text{angle}) = \frac{(\text{adj.})^2 + (\text{adj.})^2 - (\text{opp.})^2}{2 \cdot (\text{adj.}) \cdot (\text{adj.})}$$

Two Cases That the Law of Cosines Helps to Solve

The law of cosines helps to solve two cases for oblique triangles. For the first case, when two sides and an included angle are given, the law enables you to find the third side.

Sample Problem 1

Suppose that two sides of a triangle have lengths 10 centimeters and 16 centimeters and the angle between them is 60°. Find the length of the third side.

> **WORTH KNOWING**
>
> There are many proofs for the law of cosines that use methods including (but not limited to) the distance formula, the Pythagorean theorem, Ptolemy's theorem, area comparison, circle geometry, and vectors. Most of these, however, are beyond the scope of this book.

Step 1: Make a drawing of the problem situation using the data from the problem:

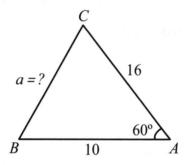

Step 2: To find the side a, use the law of cosines and substitute values from the problem:

$$a^2 = b^2 + c^2 - 2bc\cos A$$

$$a^2 = (10)^2 + (16)^2 - 2 \cdot 10 \cdot 16 \cdot \cos 60°$$

$$a^2 = 100 + 256 - 2(10)(16)\left(\frac{1}{2}\right) = 356 - 160 = 196$$

Step 3: Extract the square root to find a:

$$a = \sqrt{196} = 14$$

You find that the third side is 14 centimeters.

Solution: The third side is 14 centimeters long.

In the second case, when all three sides are given, you can find measures of all three angles.

Sample Problem 2

The lengths of a triangle's sides are 10, 20, and 24. Solve the triangle.

Step 1: Make a drawing of the problem situation using the data from the problem:

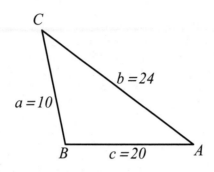

Step 2: To solve a triangle means to find its six elements (sides and angles). The problem provides you with data for the lengths of three sides, so you need to find the measures of all three angles. Let's start with $\angle A$:

$$\cos A = \frac{(\text{adj.})^2 + (\text{adj.})^2 - (\text{opp.})^2}{2 \cdot (\text{adj.}) \cdot (\text{adj.})} = \frac{20^2 + 24^2 - 10^2}{2 \cdot 20 \cdot 24} = \frac{400 + 576 - 100}{960} = \frac{876}{960} = 0.9125$$

You can find the measure of the angle with a cosine of 0.9125 by using the inverse function:

$$\cos^{-1}(0.9125) \approx 24.1°$$

$\angle A$ is approximately 24.1°.

WORD OF ADVICE

The law of cosines helps to identify acute and obtuse angles. When the cosine is positive, you have an acute angle; when the cosine is negative, you have an obtuse angle.

Step 3: Let's now find the measure of $\angle B$:

$$\cos B = \frac{(\text{adj.})^2 + (\text{adj.})^2 - (\text{opp.})^2}{2 \cdot (\text{adj.}) \cdot (\text{adj.})} = \frac{20^2 + 10^2 - 24^2}{2 \cdot 20 \cdot 10} = \frac{400 + 100 - 576}{400} = \frac{-76}{400} = -0.19$$

Because the cosine is negative, $\angle B$ is an obtuse angle. You can find the measure of the angle with a cosine of -0.19 by using the inverse function on the calculator:

$$\cos^{-1}(-0.19) \approx 101.0°$$

$\angle B$ is approximately $101.0°$.

Step 4: You can find $\angle C$ using the law of cosines as well. However, when the measures of the other two are already known, it is always easier to just use geometry to find the third angle:

$$\angle C = 180° - (101.0° + 24.1°) = 180° - 125.1° = 54.9°$$

Solution: The three angles are $101.0°$, $24.1°$, and $54.9°$.

As the wrap-up for this section, let's revisit the table from Chapter 6 and extend it by adding situations where the law of cosines can help find the missing elements of a triangle.

Given	Geometry Can Help Find	Law of Sines Can Help Find	Law of Cosines Can Help Find
ASA	Third angle	Remaining sides	
AAS	Third angle	Remaining sides	
SSA	Angle included	Angle opposite a given side and the third side	
SAS	Third angle		Third side and one angle
SSS	Third angle		Any two angles

Application of Trigonometry to Navigation and Surveying

In this section, we discuss the use of trigonometry to solve navigation and surveying problems.

The course of a plane or a ship is the angle, measured clockwise, from the north direction to the direction the plane or the ship is heading in.

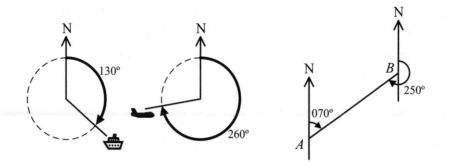

The compass bearing is a direction of which way to travel in relation to the north. It is measured in the same way as shown on the right-most picture by plotting the locations of two points and measuring the angle clockwise between a line located between them and another line pointing north. Keep in mind that this may result in large angles up to 360°.

 WORD OF ADVICE

Note that compass bearings and courses are always given with three digits, such as 050° rather than 50°.

Let's do one problem and see how it works.

Sample Problem 3

A ship proceeds on its course of 300° for 3 hours at a speed of 15 knots. Then it changes the course to 230°, continuing at 15 knots for 4 hours. At that time, how far is the ship from its starting point? (Note that 1 knot = 1 nautical mile per hour.)

Step 1: Make a drawing of the problem situation using the data from the problem:

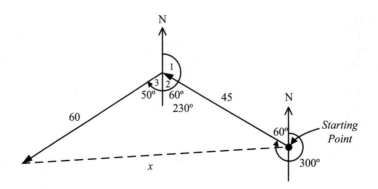

Step 2: The ship travels first along a path of length 3 · 15 = 45 nautical miles and then along a path of length 4 · 15 = 60 nautical miles.

Step 3: Let's calculate the angle between the two paths. The course is 230° = angle 1 + angle 2 + angle 3. The angle between the paths is the sum of angle 2 and angle 3. Angle 2 and angle 60° (on the first course) are alternate interior angles; therefore, angle 2 = 60°. As the figure illustrates, you can think of these two angles as mirror images of each other, hence the same measure. Angle 3 is the difference between the course angle (230°) and the sum of angles 1 and 2. Angles 1 and 2 are supplementary angles, so their sum is 180°. Thus, angle 3 = 230° − 180° = 50°. Now you can find the measure of the angle between the paths:

Angle between the paths = angle 2 + angle 3 = 60° + 50° = 110°

Step 4: Now you are ready to calculate the distance x between the starting point and the current position of the ship. You have a triangle where you know two sides (distances of different paths) and the included angle (angle between two paths), so you can apply the law of cosines to find the third side of the triangle (the needed distance):

$$x^2 = (45)^2 + (60)^2 - 2 \cdot 45 \cdot 60 \cdot \cos 110° = 2,025 + 3,600 - 5,400 \cdot (-0.3420) \approx$$
$$5,625 + 1,847 \approx 7,472$$
$$x = \sqrt{7,472} \approx 86.4$$

Solution: The distance to the starting point is 86.4 nautical miles.

Trigonometry also helps in surveying where a compass reading is usually given as an acute angle from the north-south line toward the east or west. Keep this in mind in the following examples.

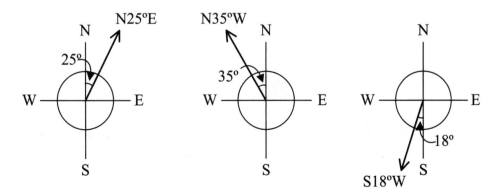

Let's now find the area of a plot of land. This is done because, very often, the owner pays an amount of tax according to the plot's area.

This problem presents you with an opportunity to hone your skills in trigonometry and apply your knowledge of both the law of sines and the law of cosines in one single problem. You'll see how the two of them support each other; one offers a shoulder to do the heavy lifting when the other is helpless. The problem also uses the trigonometric formula to calculate the area of triangles, so you have an additional chance to practice that as well.

Sample Problem 4

Find the area of a plot of land described here: From the mailbox, proceed 195 feet east along Glen Brook Road, then along a bearing of S32°E for 260 feet, then along a bearing of S68°W for 390 feet, and finally along a straight line back to the mailbox.

Step 1: Make a sketch of the problem situation using the data from the problem, and then make a drawing of the resulting quadrilateral, which as you may remember from geometry is a shape with four sides.

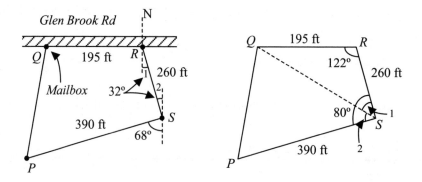

Step 2: From the given bearing on the left-side drawing, you can calculate the angles. You can deduce that:

$$\angle QRS = 90° + 32° = 122°$$

$$\angle RSP = 180° - (32° + 68°) = 80°$$

In your deduction process, you used geometric facts about the sum of supplementary angles (180°) and about the equality of alternate interior angles.

Step 3: Now that you've calculated the angles, concentrate on the area of the resulting quadrilateral. To find the area of *PQRS*, divide the quadrilateral into two triangles by the diagonal *QS*. You proceed by finding the area of each resulting triangle.

WORD OF ADVICE

When finding the area of quadrilaterals or other multilateral figures, it is always useful to divide them into triangles; their areas are found as the sum of the areas of the resulting triangles.

Step 4: You can find the area of *QRS* directly using the formula for area that involves the sine of the included angle:

$$A_{\triangle QRS} = \frac{1}{2} \cdot QR \cdot RS \cdot \sin R$$

Substitute the values:

$$A_{\triangle QRS} = \frac{1}{2} \cdot 195 \cdot 260 \cdot \sin 122° = \frac{1}{2} \cdot 195 \cdot 260 \cdot (0.8480) \approx 21,500$$

You find that the area of $\triangle QRS$ is 21,500 square feet.

Step 5: The area of $\triangle QSP$ cannot be found right away. You first need to find the length of *QS* and the measure of $\angle QSP$ (angle 2). To find *QS*, you use the law of cosines in $\triangle QRS$:

$$QS^2 = 195^2 + 260^2 - 2 \cdot 195 \cdot 260 \cdot \cos 122° = 38,025 + 67,600 - 101,400(-0.5299) = 159,400$$

Note that the sign before the third term becomes positive after you multiplied two negative factors. Thus:

$$QS = \sqrt{159,000} \approx 399$$

The length of *QS* is 399 feet.

Step 6: To find $\angle QSP$, first find the measure of $\angle RSQ$ (angle 1) using the law of sines in $\triangle QRS$:

$$\frac{\sin RSQ}{195} = \frac{\sin 122°}{399}$$

$$\sin RSQ = \frac{195 \cdot \sin 122}{399} = \frac{195 \cdot (0.8480)}{399} \approx 0.4144$$

Use the inverse function to find the measure of the angle:

$$\sin^{-1}(0.4144) \approx 24.5°$$

$\angle RSQ$ is 24.5°.

Step 7: Find the measure of $\angle QSP$:

$$\angle QSP = \angle RSP - \angle RSQ = 80° - 24.5° = 55.5°$$

$\angle QSP$ is 55.5°.

Step 8: Knowing the measures of two sides and the included angle, you can find the area of QSP:

$$A_{\triangle QSP} = \frac{1}{2} \cdot QS \cdot SP \cdot \sin 55.5° = \frac{1}{2} \cdot 399 \cdot 390 \cdot (0.8241) \approx 64{,}120$$

The area of $\triangle QSP$ is 64,120 square feet.

Step 9: As the last step, add the areas of two triangles to obtain the area of quadrilateral $QRSP$:

Area of $QRSP$ = Area of QAS + Area of QSP

Area of $QRSP$ = 21,500 + 64,120 = 85,620

The area of quadrilateral $QRSP$ is 85,620 square feet.

Solution: The area of the land plot is 85,620 square feet.

Hero's Formula for the Area of a Triangle

In Chapter 2, we mention that there are many more formulas to calculate the area of a triangle. In Chapter 6, you learned the formula that uses the sine of the included angle. But what if you don't know the measure of any of the triangles' angles? If you know three sides, then you are safe due to the wonderful formula that was discovered by another great mathematician, Hero of Alexandria.

WORTH KNOWING

Hero (or Heron) of Alexandria (10–70 C.E.) was a great mathematician and engineer who lived in Alexandria, Roman Egypt. He is considered the greatest experimenter of ancient times. Among his most famous inventions was a wind wheel—the earliest known example of using a wind-driven wheel to power machinery.

Even though the proof of this formula is not difficult to understand, it is lengthy, so we are not going to prove it here. We will just explain how to use it.

Suppose you have a triangle with sides a, b, and c. Let's agree that you denote by s the following expression:

$$s = \frac{(a+b+c)}{2}$$

Then, the area of a triangle can be calculated by Hero's formula:

$$A = \sqrt{s(s-a)(s-b)(s-c)}$$

Let's solve the last problem of this chapter using this formula.

Sample Problem 5

Find the area of triangle with sides 5, 12, and 13.

Step 1: It is always useful to find the value for s separately and then substitute it into the formula:

$$s = \frac{(a+b+c)}{2} = \frac{5+12+13}{2} = 15$$

You find that s is 15.

Step 2: Substitute the values for triangle sides and for s into Hero's formula:

$$A = \sqrt{s(s-a)(s-b)(s-c)} = \sqrt{15(15-5)(15-12)(15-13)} = \sqrt{15 \cdot 10 \cdot 3 \cdot 2} = \sqrt{900} = 30$$

Solution: The area of the triangle is 30 square units.

Practice Problems

Problem 1: In $\triangle ABC$, $\angle C$ measures 60°, a is 2, and b is 8. Find the length of side c.

Problem 2: A triangle has sides of length 10, 16, and 20. Find the measure of its largest angle.

Problem 3: After leaving an airport, a plane flies for 1 hour at a speed of 250 kilometers per hour on a course of 200°. Then it flies for 3 hours on a course of 340° at a speed of 200 kilometers per hour. At this time, how far from the airport is the plane?

Problem 4: Sketch the following plot and find its area: from the southeast corner of the local park on Cherry Hill Road, proceed S78°W for 250 meters along the southern boundary of the park to the old oak tree. From the tree, proceed S15°E for 180 meters to Mulberry Lane, then N78°E along Mulberry Lane until it intersects Cherry Hill Road, and finally N30°E along Cherry Hill Road back to the starting point.

Problem 5: Find the area of a triangle with sides of 5, 6, and 7 using Hero's formula.

The Least You Need to Know

- In $\triangle ABC$, if $\angle C$ is an acute angle, then $c^2 = a^2 + b^2 - 2ac\cos C$. This is the law of cosines.
- Dividing multilaterals into triangles helps to find those figures' areas.
- The cosine of an obtuse angle is the cosine of its supplement multiplied by –1.
- Hero's formula for a triangle's area is as follows:

$$A = \sqrt{s\,(s-a)\,(s-b)\,(s-c)}, \text{ where } s = \frac{(a+b+c)}{2}$$

Trigonometric Functions and the Unit Circle

Part

3

This part is all about angles. You learn how to measure angles using degrees and radians, and formulas for conversion from degree measure to radians and vice versa are introduced.

You explore coterminal angles and learn why one angle can have many names—actually, an infinite number of names.

Then, you learn about the unit circle and trigonometric functions for all angles—positive and negative—ranging in their measure from negative infinity to positive infinity. Introduction of reference angles will provide a powerful and easy way to calculate values of trigonometric functions for any angle.

Finally, we discuss the domain and period of trigonometric functions and determine which trigonometric functions are even and which ones are odd.

Angles and Rotations

Chapter 8

In This Chapter

- Finding different measures for one angle
- Understanding coterminal angles and how to work with them
- Converting from degrees to radians and vice versa
- Estimating the heights or sizes of distant objects

In previous chapters, we explored the first historical perspective of trigonometric functions and identified these functions using right triangles. Another way of studying trigonometric functions—we'll call it another perspective—deals with the unit circle. Before we identify trigonometric functions using the unit circle, let's discuss rotations, angles, and the two ways of measuring angles.

Measuring Angles

So far we have mostly discussed acute angles and their trigonometric functions. We also touched on obtuse angles and some of their sines and cosines. But at this point, you might be wondering how to determine the sine of 500° or the cosine of 1,000°.

What do these large measures of degrees measure? Certainly, they cannot measure any of the angles in a triangle, because these can be only between 0° and 180° (even though nobody has ever seen an angle in a triangle with a measure of 180°). They cannot measure any central angles or arcs in a circle, either, because these can be only between 0° and 360°.

Imagine yourself standing in front of a carousel that runs around a circle. Suppose also that your younger sibling sits on a toy horse that is a part of the carousel. The following figure shows its circular track.

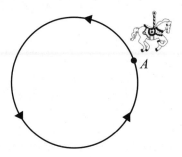

If the horse starts at point A, travels around a full circle, and comes back at point A, we say that it has made one full rotation (or revolution) around the circle. Because a circle is divided into 360 degrees, we can say that the horse has rotated around the circle by 360°. Certainly, your younger sibling will not be happy to stop after just one rotation, so the carousel continues past point A and travels the circle again, so that it has rotated around the circle twice—360° + 360° = 720°. When it makes three full rotations, it already has rotated more than 1,000° (360° · 3 = 1,080°).

Consider another example. The hour hand of a clock makes a full rotation of 360° every 12 hours. But in contrast to the carousel, which rotates counterclockwise, the hour hand rotates clockwise. In mathematics, it is important to distinguish between these two directions of rotation: a counterclockwise rotation is considered to be positive, and the clockwise rotation is negative. So we can say that in 24 hours (one full day), the hour hand of a clock performs a rotation of –720°.

The following figure shows different rotations.

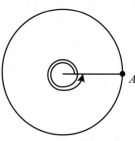
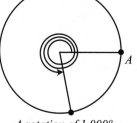
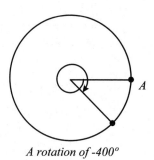

A rotation of 720°   *A rotation of 1,000°*   *A rotation of -400°*

Degree Measures

Remember that a common unit for measuring angles is the degree and there are 360 degrees in one rotation.

WORTH KNOWING

The custom of having 360 degrees in one rotation can be traced back to the Babylonian civilization and its numerical system that was based on the number 60. As discussed in previous chapters, it is a convenient number that can be evenly divided by many other numbers.

Angles can be measured more precisely by dividing 1 degree into 60 minutes, and further dividing 1 minute into 60 seconds. For instance, an angle 35 degrees, 18 minutes, and 5 seconds can be written as 35°18'5". Converting such angles into decimals entails the following:

$$35°18'5" = 35 + \frac{18}{60} + \frac{5}{60 \cdot 60} = 35 + 0.3 + \frac{5}{3,600} \approx 35 + 0.3 + 0.00139 \approx 35.30139°$$

In many fields of science and engineering, angles are measured by using the same angle measure (degrees), but the angles can be stated in different positions or places. Recall the examples with surveying and navigation, where angles are measured from the direction of the north and such is the standard format of expressing navigational directions. But in trigonometry, we have to draw angles in a standard position (where the angle "starts" on the x-axis) so that everyone can be assured that we are discussing the same thing.

Coterminal Angles

When an angle is shown in a coordinate plane, it usually appears in the standard position, with its vertex at the origin and its *initial ray* (starting side) along the positive x-axis. Its *terminal ray* (ending side) moves counterclockwise from the initial side. If the terminal side moves clockwise, then the angle has a negative measure.

DEFINITION

Rays are portions of a line that start at a point and go off in a particular direction to infinity. An **initial ray** is a ray that defines one side of an angle, traditionally one from which the angle measure starts. A **terminal ray** is the second ray of an angle that is located where the angle measure ends.

Note that only the direction of the terminal ray determines the angle; the lengths of its sides do not affect its degree measure. You can extend the angle's rays as long as you want, but the angle's measure will always stay the same.

Consider two angles: 290° and –70°. They have the same terminal side but we refer to them differently. These two angles are called *coterminal angles*.

 DEFINITION

Two angles are called **coterminal angles** if they have the same terminal side.

The following figure shows angles with both positive and negative measures that are coterminal angles.

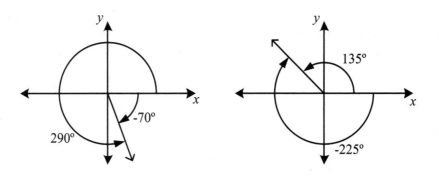

There is an infinite number of ways to provide an angle measure for a specific terminal ray. Notice that the initial angle of 50° on the following figure is coterminal to the angle measuring 410° (one full rotation plus initial 50°); it is also coterminal with the angle measuring 770° (two full rotations plus initial angle), and so forth.

Because rotation can also go clockwise, the initial angle of 50° is coterminal with angles –310°, –670°, and so forth.

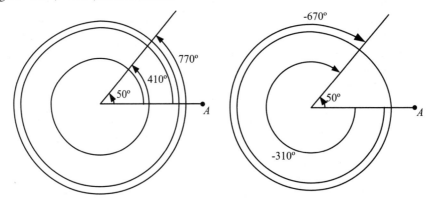

This means that one angle can be identified by many measures. Which one should we choose? The common practice is to call all angles with numbers that have an absolute value less than 180°. For example, it is preferable to identify an angle as –25° rather than 335°. Likewise, expressing an angle as –178° is preferred to stating it as 182°, even though both of them are very close to 180°.

To summarize: in order to obtain the measures of coterminal angles, you either need to add or subtract rotations over and over again. This reduces measures of large angles to the smaller coterminal angles that are much more convenient to work with.

If you choose the initial angle θ, then this angle is coterminal with all the following angles:

$$\theta \rightarrow (\theta + 360°) \rightarrow (\theta + 720°) \rightarrow (\theta + 1{,}080°) \rightarrow \ldots$$

$$\theta \rightarrow (\theta - 360°) \rightarrow (\theta - 720°) \rightarrow (\theta - 1{,}080°) \rightarrow \ldots$$

Let's consider several examples and try to rename angles 900° and –1,150° by using measures that are between 0° and 180°:

$$900° \rightarrow 900° - 360° \rightarrow 540° - 360° \rightarrow 180°$$

Thus, an angle measuring 900° is coterminal with an angle of 180°.

$$-1{,}050° \rightarrow -1{,}050° + 360° \rightarrow -690° + 360° \rightarrow -330° + 360° \rightarrow 30°$$

Thus, an angle measuring –1,050° is coterminal with an angle of 30°.

Radian Measure

Relatively recently in mathematical history, another unit of angle measure has come into common use: the *radian*. With the degree measure, the rule of dividing a circle into 360 degrees comes from history rather than pure mathematics, which makes a degree an arbitrary unit for measuring angles. It turns out that radian measure is a more convenient way to measure angles because radian measure does not depend on any arbitrary unit.

DEFINITION

The **radian** measure of an angle is the ratio of the arc (part of the circumference) that the angle traces to the radius of any circle with a center at the vertex of the angle.

When an arc of a circle has the same length as the radius of the circle, the measure of the central angle is, by definition, one radian.

To measure an angle in radians, you place its vertex at the center of any circle and consider the length of arc *AB*, as measured in any unit of length. Certainly, the length of this arc depends on the size of ∠*AOB*. But it also depends on the size of the radius of the circle as the following figure illustrates:

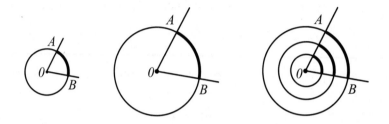

This means that you cannot simply take the measure of this arc as the measure of the angle. Both a very large arc and a small arc (from a large and a small circle, respectively), could stem from the same angle measure. Meanwhile, the ratio of the length of the arc to the radius of the circle depends only on the size of the angle.

The situation with radian angle measures resembles the definition of the trigonometric functions in a right triangle. For example, the sine of an angle is the ratio between an opposite side and the hypotenuse, and this ratio does not depend on the lengths of triangle's sides.

In general, the radian measure of the central ∠*AOB* is the number of radius units in the length of arc *AB*. This accounts for the name of this angle measure unit: radian.

Now let's work with radians to find the radian measure of some angles.

Sample 1: Find the radian measure of an angle of 60°.

Step 1: Place the vertex of this angle at the center of a circle of radius *r*, and examine the length of the arc formed in between the rays of the angle.

Step 2: Because $\dfrac{60}{360} = \dfrac{1}{6}$, this arc is one sixth of the circumference of the circle.

Step 3: The circumference C is found by the formula $C = 2\pi r$. Divide it by 6 to find the length of the arc: $\dfrac{2\pi r}{6} = \dfrac{\pi r}{3}$.

Step 4: By definition, the radian measure is the ratio of the length of the arc $(\dfrac{\pi r}{3})$ to the radius. Thus, the radian measure of 60° is $\dfrac{\frac{\pi r}{3}}{r} = \dfrac{\pi}{3}$.

Sample 2: Find the radian measure of an angle of 45°.

Place the vertex of this angle at the center of a circle of radius r, and examine the arc it forms along the circumference. Because $\dfrac{45}{360} = \dfrac{1}{8}$, this arc is one eighth of the circumference of the circle. The circumference C is found by the formula $C = 2\pi r$.

Divide it by 8 to find the length of the arc: $\dfrac{2\pi r}{8} = \dfrac{\pi r}{4}$. By definition, the radian measure is the ratio of the length of the arc $(\dfrac{\pi r}{4})$ to the radius. Thus, the radian measure of 45° is $\dfrac{\frac{\pi r}{4}}{r} = \dfrac{\pi}{4}$.

Sample 3: Let's find out how many radians are in one full rotation (360°). You need to divide the whole circumference by r: $2\pi r \div r = 2\pi$. Thus, one rotation or 360° = 2π radians, or π radians = 180°.

This discussion provides two conversion formulas between degrees and radians.

$$1 \text{ radian} = \frac{180}{\pi} \text{ degrees} \approx 57.2957 \text{ degrees}$$

$$180° = \pi \text{ radians} \qquad\qquad 360° = 2\pi \text{ radians}$$

WORTH KNOWING

It is common for radian angle measures to be expressed without any symbols or units next to them. While not all radian measures are stated in terms of π, many of the most common angles are. The presence of π is thus a good clue to mean that the angle measure is stated in radians.

Let's use the conversion formulas and see how they work. We use the approximate value of 3.14 for π. Let's first convert 186° into radians and then convert 1.45 radians into degrees:

$$186° = 186 \cdot \frac{\pi}{180} \approx 3.24 \text{ radians}$$

$$1.45 \text{ radians} = 1.45 \cdot \frac{180}{\pi} \approx 83.1°$$

Angle measures that can be expressed evenly in degrees cannot be expressed evenly in radians and vice versa. Therefore, angles measured in radians are often given as fractional multiples of π.

Angles with measures that are multiples of $\frac{\pi}{4}$, $\frac{\pi}{3}$, and $\frac{\pi}{6}$ frequently appear in trigonometry. The following figure presents the most commonly used angles.

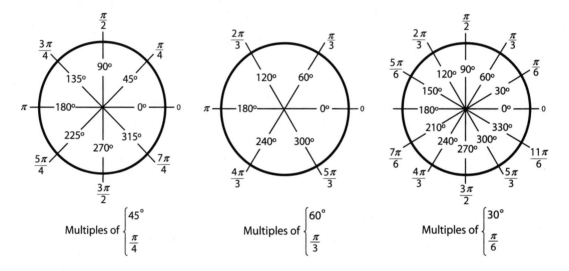

Now apply your knowledge of angles and their measures to solve the following problem.

Sample Problem 1

Find two angles, one positive and one negative, that are coterminal with the angle $\frac{\pi}{3}$. Sketch all three angles.

Step 1: To find a positive coterminal angle, you need to add 360° degrees or 2π radians (full rotation):

$$\frac{\pi}{3} + 2\pi = \frac{\pi}{3} + \frac{6\pi}{3} = \frac{7\pi}{3}$$

Step 2: To find a negative coterminal angle, you need to subtract 360° degrees or 2π radians (full rotation):

$$\frac{\pi}{3} - 2\pi = \frac{\pi}{3} - \frac{6\pi}{3} = -\frac{5\pi}{3}$$

Step 3: Sketch all three angles:

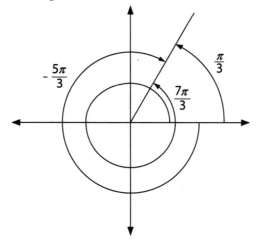

Sectors of Circles

In this section, you learn to find the arc length and area of a *sector* of a circle and tackle problems involving apparent size. But before you do so, you need to be familiar with some related terminology and formulas.

DEFINITION

A **sector** of a circle is the region bounded by a central angle and the intercepted arc.

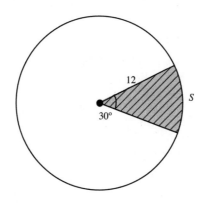

In the preceding figure, the sector of the circle is the shaded region. It is clear that the length of the arc is some fraction of the circle's circumference and the sector's area is some fraction of the circle's area.

Suppose that the central angle of a sector is 30° and its radius is 12. Then the arc length of the sector is $\dfrac{1}{12}$ of the whole circumference because 360° ÷ 30° = 12. The whole circumference is $2\pi r$, and the arc length can be calculated using this fact and by substituting the value of 12 for the radius: $\dfrac{1}{12} \cdot (2\pi r) = \dfrac{2\pi \cdot 12}{12} = 2\pi$

The area of the sector can be calculated similarly:

$$\frac{1}{12}\left(\pi r^2\right) = \frac{1}{12} \cdot \pi\left(12^2\right) = 12\pi$$

The following are formulas for the arc length s and the area K of a sector with central angle θ.

If the angle θ is given in degrees, then:

$$s = \frac{\theta}{360} \cdot 2\pi r \qquad\qquad K = \frac{\theta}{360} \cdot 2\pi r^2$$

If the angle θ is given in radians, then:

$$s = r\theta \qquad\qquad K = \frac{1}{2}r^2\theta \text{ or } K = \frac{1}{2}rs$$

Note that we obtained formula $K = \dfrac{1}{2}rs$ by combining formulas $s = r\theta$ and $K = \dfrac{1}{2}r^2\theta$ as follows:

$$K = \frac{1}{2}r^2\theta = \frac{1}{2}r\left(r\theta\right) = \frac{1}{2}rs$$

You are now ready to try your knowledge in solving problems. Let's start with an easy one that requires only the use of formulas.

Sample Problem 2

A sector of a circle has an arc length of 8 centimeters and an area of 80 square centimeters. Find the circle's radius and the measure of the arc's central angle.

Step 1: You know the arc length and the sector's area, so you can use one of the formulas stated previously to find the radius:

$$K = \frac{1}{2}rs$$

Substitute values for the arc length and the area:

$$80 = \frac{1}{2}r \cdot 8 = 4r$$

Divide both sides by 4:

$$r = 20$$

The radius is 20 centimeters.

Step 2: To find the angle, use the following formula:

$$s = r\theta$$

Substitute values for the arc length from the problem and the value for r that you just found:

$$8 = 20 \cdot \theta$$

$$\theta = \frac{8}{20} \approx 0.4$$

The central angle is 0.4 radians.

Step 3: To find the angle's measure in degrees, use the conversion formula:

$$0.4 \cdot \frac{180}{\pi} \approx \frac{0.4 \cdot 180}{3.14} \approx 23°$$

Solution: The radius of the circle is 20 centimeters and the central angle is 23°.

Another common trigonometry problem you may encounter asks you to find the distance traveled by points on a car wheel, by a potter's wheel, by the tip of a clock's hour (or minute) hand, or by a Ferris wheel. All of these problems use the concepts you just learned and can be approached with the same reasoning, provided that the distance traveled is just the length of the arc that is cut off by a specific central angle.

Suppose you have a wheel with a radius of 1 foot and it rolls without slipping along a straight road. Let's discuss the distance a point A travels.

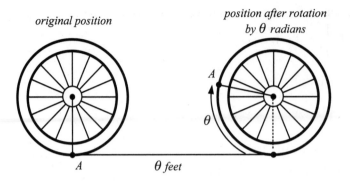

original position

position after rotation by θ radians

A

θ

A

θ feet

The distance that point A travels, in feet, is just the length of the arc that the angle θ cuts off. Because the radius is just 1 foot, the length of the arc can be found as $s = r\theta = 1 \cdot \theta = \theta$ feet.

When the radius of a circle is 1 length unit and the wheel makes a full rotation, the distance traveled is 2π (because the full angle is 360° or 2π). When a wheel of radius 1 length unit rotates through half a rotation, the distance traveled is just π because the central angle in this case is 180° (or π in radians). Keeping this all in mind, let's do the following problem.

Sample Problem 3

An artist uses a potter's wheel to make a clay vase. The diameter of the wheel is 12 inches and it spins at 120 revolutions per minute. Find the distance traveled in 2 minutes by a point on the outer edge of the wheel.

Step 1: In 1 minute, the wheel does 120 rotations, so in 2 minutes it will make:

120 · 2 = 240 rotations

Step 2: Find the radius knowing that the diameter is 12 inches:

12 ÷ 2 = 6 inches

Step 3: Find the distance traveled by the point of the wheel after one rotation (in this case, the length of the arc is equal to the whole circumference):

$s = r\theta = 2\pi r = 2\pi \cdot 6 = 12\pi$

Step 4: By multiplying the distance traveled after one rotation by 240 rotations, you find the whole distance D traveled:

$$D = 12\pi \cdot 240 \approx 12 \cdot 3.14 \cdot 240 \approx 9,043 \text{ inches}$$

Because there are 12 inches in 1 foot, divide 9,043 by 12 to find this distance in feet:

$$9,043 \div 12 \approx 753.583 \text{ feet}$$

Step 5: Let's convert the fractions of the feet into inches. We can take the known length in inches and subtract from it the number of inches that comprise 753 feet. Thus, $9,043 - 753 \cdot 12 = 7$ inches.

Solution: The distance traveled by the point is 753 feet and 7 inches.

Let's do a similar problem that involves the minute hand of a clock.

Sample Problem 4

The minute hand of the City Hall clock in Brooksville is 9 feet. How far does the tip of the minute hand travel from 4:10 P.M. until 4:30 P.M.?

Step 1: Calculate the angle that the tip of the minute hand travels. From 4:10 until 4:30, there are 20 minutes, and 20 minutes is $\frac{1}{3}$ of an hour. That means that the minute hand will travel $\frac{1}{3}$ of the full circle (360°), which translates into 120°.

 WORD OF ADVICE

It is always more convenient to use the radian measure of angles when solving the types of problems that are discussed in this chapter.

Step 2: Convert degrees into radians. Because the full circle is 2π, one third (120°) of it is $\frac{2\pi}{3}$. You can also use the conversion formula to get the same result:

$$120° \rightarrow 120 \cdot \frac{\pi}{180} = \frac{2\pi}{3}$$

Step 3: Knowing the angle measure that you just calculated and the length of the hand, you can use the formula for the length of the arc:

$$s = \theta r = \frac{2\pi}{3} \cdot 9 = \frac{18\pi}{3} = 6\pi \approx 18.8$$

The length of the arc is about 18.8 ft.

Solution: The minute hand traveled about 18.8 feet.

The last type of problem we are going to discuss is related to the apparent size of visible objects. You know that the sun is much larger than the moon, but when looking at them, you perceive them as being much closer in size than they really are—the sun is thousands of times bigger than the moon, but it may seem only twice as large in the morning when both are in the sky. This happens because the sun is much farther away than the moon.

Thus, how big an object looks depends not only on its size but also on the angle between the edges of the object that a person sighting the object can measure from the point where they are observing it from. The measure of this angle is called the object's apparent size.

For example, Jupiter is $8 \cdot 10^8$ km from the Earth and it has an apparent size of $0.01°$ as shown on the following figure.

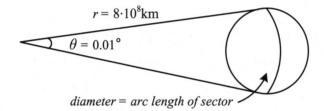

Let's do the last problem in this chapter that discusses the apparent size of objects.

Sample Problem 5

Sam travels in a car toward Spooky Mountain at a speed of 80 kilometers per hour. The apparent size of the mountain is $0.5°$. Fifteen minutes later the apparent size of the same mountain is $1°$. How tall is Spooky Mountain?

Step 1: Let's denote the height of the mountain by s and the distance between the second point of observation and the mountain by x.

Step 2: Because the car traveled 15 minutes between the two points of observation, it traveled 20 kilometers (since its speed is 80 kilometers per hour). Using the variables and the data from the problem, let's sketch the situation that the problem describes:

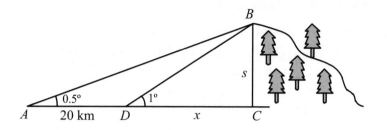

Step 3: Convert degrees into radians:

$$1° \rightarrow \frac{1 \cdot \pi}{180} = \frac{\pi}{180} \text{ radians}$$

$$0.5° \rightarrow \frac{0.5 \cdot \pi}{180} = \frac{1 \cdot \pi}{2 \cdot 180} = \frac{\pi}{360} \text{ radians}$$

WORD OF ADVICE

When calculating the heights or diameters of objects located far away, you can approximate and consider these heights or diameters as being the arc lengths of a sector.

Step 4: You can consider CDB as a sector of a circle with central angle 1° ($\frac{\pi}{180}$ radians) and the radius of x. The height of Spooky Mountain, BC (denoted as s), is the arc of this circle that can be calculated using the formula for the arc:

$$s = \theta r = \frac{\pi}{180} \cdot x$$

Let's express x though s because you need it for upcoming steps of this problem, where we will use the arc length formula to find the height of the mountain.

Cross-multiply the previous equation:

$$180 \cdot s = \pi \cdot x$$

$$x = \frac{180 \cdot s}{\pi}$$

Step 5: Similarly, you can consider ABC as a sector of a circle with a central angle of $0.5°$ ($\dfrac{\pi}{360}$ radians) and the radius of $x + 20$, because the distance between A and D is 20 kilometers. Then, the height of Spooky Mountain, BC (that is denoted as s), is the arc of this circle that can be calculated using the formula for the arc:

$$s = \theta r = \frac{\pi}{360} \cdot (x + 20)$$

Step 6: Substitute expression for x ($x = \dfrac{180 \cdot s}{\pi}$) from Step 4 into the last equation to eliminate two unknowns and simplify:

$$s = \theta r = \frac{\pi}{360} \cdot (x + 20) = \frac{\pi}{360} \left(\frac{180 \cdot s}{\pi} + 20 \right)$$

$$s = \frac{\pi}{360} \left(\frac{180 \cdot s}{\pi} + 20 \right) = \frac{\cancel{\pi} \cdot \cancel{180}^{\,1} \cdot s}{_{2}\cancel{360} \cdot \cancel{\pi}} + \frac{\pi \cdot \cancel{20}^{\,1}}{_{18}\cancel{360}} \text{ thus,}$$

$$s = \frac{s}{2} + \frac{\pi}{18}$$

Step 7: To solve the fractional equations, it is always useful to multiply each term by the least common denominator (LCD) to eliminate fractions. In this case, the LCD is 18:

$$18 \cdot s = \frac{\cancel{18}^{\,9} \cdot s}{\cancel{2}_{1}} + \frac{\cancel{18}^{\,1} \cdot \pi}{\cancel{18}_{1}}$$

$$18s = 9s + \pi$$

$$9s = \pi \text{ or } s = \frac{\pi}{9} \approx \frac{3.14}{9} \approx 0.350$$

The height of the mountain is approximately 0.350 kilometers or 350 meters (1 kilometer = 1,000 meters).

Solution: The height of Spooky Mountain is 350 meters.

In this chapter, you were introduced to the definition of the radian, and shown how to convert angle measures between radians and degrees. Both units have sets of applications where each is convenient and knowledge of both helps you apply the proper unit for each application. You have also learned about coterminal angles and the ability to reduce large (and sometimes scary-looking) angles into their much smaller equivalents, which are much more convenient to work with. Last but not least, you have also learned how to measure heights of distant objects utilizing trigonometry.

Practice Problems

Problem 1: Convert each degree measure to radians: 180°, 240°, 135°, 225°, and –135°. Leave answers in terms of π.

Problem 2: Convert each radian measure to degrees: $\dfrac{3\pi}{4}, \dfrac{11\pi}{6}, -\dfrac{7\pi}{6}, \dfrac{2\pi}{3}$, and $\dfrac{5\pi}{6}$.

Problem 3: Find two angles, one positive and one negative, that are coterminal with each given angle: 400° and $\dfrac{\pi}{4}$.

Problem 4: A race car moves around a circular track. Find the speed of the car if the track has a radius of 250 meters and the car covers a quarter of the track in 5 seconds.

Problem 5: The diameter of the moon is approximately 3,500 kilometers. Its apparent size from the Earth is 0.0087 radians. How far is the moon from Earth?

The Least You Need to Know

- Each angle in a standard position has an infinite number of positive and negative coterminal angles.
- To convert degrees to radians, use the following formula:

  $$1 \text{ degree} = \frac{\pi}{180} \text{ radians}$$

- To convert radians to degrees, use the following formula:

  $$1 \text{ radian} = \frac{180}{\pi} \text{ degrees}$$

- The formulas for the arc length s and area K of a sector with central angle θ (in radians) are $s = r\theta$ and $K = \dfrac{1}{2}r^2\theta$.

The Unit Circle Approach

In This Chapter

- Coordinates and the unit circle
- Understanding how the unit circle works
- Calculating the trig functions for large angles
- Deciding whether the sine of 100,000° is positive or negative

In this chapter, we use the second historical perspective to present trigonometric functions. As emphasized several times already, the first perspective through right triangles is a natural way to introduce trigonometric functions, and it proves to be useful for mathematicians, astronomers, and engineers. It allows them to calculate distances to remote objects, find the heights of tall objects, predict eclipses, find areas of vast plots of land, and fulfill many other essential tasks. But all this comes with a major impediment—this first approach through right triangles does not enable us to consider trigonometric functions of all angles. This is where the second approach steps in to remove these limitations. We are now going to expand the trigonometric functions to angles of any measure, positive and negative, and explore their properties. We start our exploration with the unit circle.

What Is the Unit Circle?

The *unit circle* is the basic and convenient tool for describing all possible angles from 0° to 360°, including all their multiples and all the negative angles. In other words, all positive and negative angles with measures from negative infinity to positive infinity can be explored using the unit circle. Isn't this a powerful tool?

But what is the unit circle? It is time to meet it. The unit circle is a circle with its center at the origin of the coordinate plane and with a radius of 1 unit—which is how the circle got its name. The equation of any circle with its center at the origin is $x^2 + y^2 = r$, where r is the radius of the circle. Because the radius of the unit circle is 1, its equation is $x^2 + y^2 = 1$. The coordinates for the points lying on the unit circle and intersecting the axes are (1,0), (0,1), (–1,0), and (0,–1).

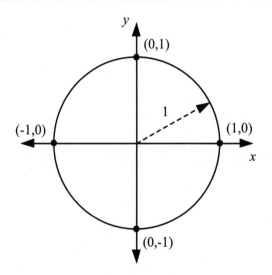

Suppose $P(x,y)$ is a point on the unit circle $x^2 + y^2 = 1$, and θ is an angle in standard position with terminal ray OP. Let's drop the altitude from point P to the axis x.

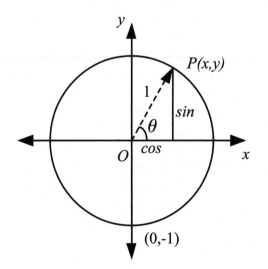

As noted in the figure, $\sin\theta$ and $\cos\theta$ are the y- and x-coordinates of the point P:

$$\sin\theta = \frac{y}{r} = \frac{y}{1} = y \qquad\qquad \cos\theta = \frac{x}{r} = \frac{x}{1} = x$$

Then you can define all six trigonometric functions as:

$$\sin\theta = y \qquad\qquad\qquad \csc\theta = \frac{1}{y}, y \ne 0$$

$$\cos\theta = x \qquad\qquad\qquad \sec\theta = \frac{1}{x}, x \ne 0$$

$$\tan\theta = \frac{y}{x}, x \ne o \qquad\qquad \cot\theta = \frac{x}{y}, y \ne 0$$

Note that the functions in the right column are the reciprocals of the corresponding functions on the left.

The tangent and the secant are not defined, meaning that they do not exist as finite values, when $x = 0$. For example, because the angle 90° or $\frac{\pi}{2}$ corresponds to $(x,y) = (0,1)$, tan90° and sec90° are undefined. Similarly, the cotangent and the cosecant are undefined when $y = 0$. For example, because angle 0° corresponds to $(x,y) = (1,0)$, cot0° and csc0° are undefined.

Recall from Chapter 3 that we divide the coordinate plane into four quadrants. Let's consider in which quadrants the trigonometric functions have positive and negative values. For example, the sine depends only on the value of y; therefore, the sine function has positive values for all angles in the first and second quadrants. The value of the cosine function is defined by the value x only; therefore, it is positive for all angles in the first quadrant and negative for all angles in the second quadrant.

Reasoning along these lines, you can decide whether other trigonometric functions are positive or negative. This is summarized in the following figure.

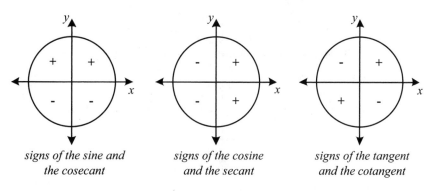

signs of the sine and   signs of the cosine   signs of the tangent
the cosecant   and the secant   and the cotangent

Placing Points on the Unit Circle

This section identifies some important angles and provides values of x and y for them. Let's start with *quadrantal angles*. In degrees, these angles are 0°, 90°, 180°, 270°, and all other angles that are coterminal with them.

📖 **DEFINITION**

A **quadrantal angle** is an angle that terminates on the x- or y-axis.

In radians, these angles are 0, $\dfrac{\pi}{2}$, π, $\dfrac{3\pi}{2}$, and all other angles coterminal with them.

The coordinates of the quadrant angles on the unit circle can be only 0, 1, or –1. But coordinates x and y can be determined for other commonly used angles.

The following figure shows the (x,y) coordinates for these angles. Note that the first coordinate represents the value of the cosine and the second coordinate represents the value for the sine.

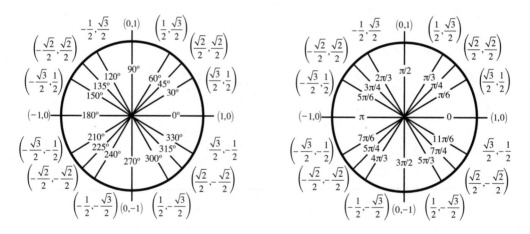

Any combination of these two coordinates corresponds to a different point on the unit circle, and therefore they should satisfy the unit circle equation. For example, let's take coordinates $\left(\dfrac{1}{2}, -\dfrac{\sqrt{3}}{2}\right)$, corresponding to $\dfrac{5\pi}{3}$ angle, and work them into the circle's equation:

$$x^2 + y^2 = \left(\frac{1}{2}\right)^2 + \left(-\frac{\sqrt{3}}{2}\right)^2 = \frac{1}{4} + \frac{3}{4} = 1$$

It works. All other coordinates work the same way. We also want to emphasize that the points we placed on the unit circle in the preceding figure are the most frequently used points when solving trigonometric problems, but they certainly don't exhaust all possible points because there is an infinite number of points on the unit circle.

WORD OF ADVICE

Often, when working with the coordinates on the unit circle and trying to find the values of trigonometric functions, you need to work with radicals and complex fractions. The generally accepted mathematical convention is to avoid leaving a square root in the denominator of a fraction. Here are some key rules:

$\sqrt{2} \cdot \sqrt{2} = 2$, $\sqrt{3} \cdot \sqrt{3} = 3$, and so forth

$\dfrac{1}{\sqrt{2}} = $ rationalize the denominator $= \dfrac{1 \cdot \sqrt{2}}{\sqrt{2} \cdot \sqrt{2}} = \dfrac{\sqrt{2}}{2}$

$\dfrac{\frac{a}{b}}{\frac{c}{d}} = \dfrac{a \cdot d}{b \cdot c}$

In the next section, we discuss angles on the unit circle in more detail.

Making Sense of the Unit Circle

For many, the unit circle looks intimidating—too many angles and an overwhelming number of radicals. It seems that no one can remember it. Following are some guidelines for how to understand and remember the unit circle.

- **Guideline 1:** The first thing to remember is that there are four quadrant angles (0° and 360° being the same angle). Their sines and cosines can be 0, 1, or –1, as previously mentioned. These angles are:

0°	90°	180°	270°	360°
0	$\dfrac{\pi}{2}$	π	$\dfrac{3\pi}{2}$	2π
(1,0)	(0,1)	(–1,0)	(0,–1)	(1,0)

- **Guideline 2:** The second thing to remember is that there are three angles in the first quadrant with sines and cosines you need to memorize:

30°	45°	60°
$\dfrac{\pi}{6}$	$\dfrac{\pi}{4}$	$\dfrac{\pi}{3}$
$\left(\dfrac{\sqrt{3}}{2}, \dfrac{1}{2}\right)$	$\left(\dfrac{\sqrt{2}}{2}, \dfrac{\sqrt{2}}{2}\right)$	$\left(\dfrac{1}{2}, \dfrac{\sqrt{3}}{2}\right)$

Note that all coordinates are fractions with the denominator of 2 and all numerators are radicals (even 1 is technically a radical because $1 = \sqrt{1}$). Recall that the second coordinate always stands for the sine since it is on the y-axis (as discussed in the beginning of this chapter), so the numerators follow this simple pattern from the smallest angle to the largest: for sines: $\sqrt{1}$, $\sqrt{2}$, $\sqrt{3}$; for cosines, it goes in an opposite order: $\sqrt{3}$, $\sqrt{2}$, $\sqrt{1}$.

After discussing these two guidelines, you are now done with 7 angles on the unit circle (4 quadrants and 3 special angles). You are left with 9 more.

- **Guideline 3:** Each special angle from the first quadrant has 3 multiples, one in each of the three other quadrants; therefore, there are 9 of them, since $3 \cdot 3 = 9$. These multiples have the same values for sines and cosines as their corresponding special angle, so you don't have to remember any new values; you just need to watch the signs of the trig functions in each quadrant.

Let's first list these multiples for each special angle and then see how you identify their functions' values.

Start with multiples of $\dfrac{\pi}{6}$ (30°):

$$\frac{\pi}{6}, \frac{2\pi}{6}, \frac{3\pi}{6}, \frac{4\pi}{6}, \frac{5\pi}{6}, \frac{6\pi}{6}, \frac{7\pi}{6}, \frac{8\pi}{6}, \text{ and so forth up to } \frac{12\pi}{6} = 2\pi$$

Some multiples can be simplified like $\dfrac{4\pi}{6} = \dfrac{2\pi}{3}$, which becomes a multiple of another special angle. However, you are interested only in the multiples that cannot be simplified in that manner. There are only three multiples that cannot be reduced: $\dfrac{5\pi}{6}$, $\dfrac{7\pi}{6}$, and $\dfrac{11\pi}{6}$. You can find them on the unit circle.

Their sine and cosine values are the same as for $\dfrac{\pi}{6}$ except for the signs. Let's identify the sine and the cosine for $\dfrac{11\pi}{6}$. Because this angle is in the fourth quadrant, its sine (y-coordinate) is negative and its cosine (x-coordinate) is positive; therefore, $\sin\dfrac{11\pi}{6} = -\dfrac{1}{2}$ and $\cos\dfrac{11\pi}{6} = \dfrac{\sqrt{3}}{2}$.

Multiples of other special angles can be found by reasoning along the same lines. They are:

$$\frac{\pi}{4} : \frac{3\pi}{4}, \frac{5\pi}{4}, \frac{7\pi}{4}$$

$$\frac{\pi}{3} : \frac{2\pi}{3}, \frac{4\pi}{3}, \frac{7\pi}{3}$$

We hope that this dissection of the unit circle helped you understand how it works and provided useful guidelines on how to remember the unit circle angles and their trigonometric functions' values.

Evaluating Trigonometric Functions of Real Numbers

The points that we placed on the unit circle help us to find trigonometric functions of some often-used and not-so-convenient angles.

Sample Problem 1

Evaluate the six trigonometric functions for angle $\dfrac{7\pi}{6}$.

Step 1: Find the angle on the unit circle and record its coordinates:

$$\left(-\frac{\sqrt{3}}{2}, -\frac{1}{2}\right)$$

Step 2: Find three main trigonometric functions:

Because the second coordinate is the sine and the first coordinate is the cosine, you obtain:

$$\sin\frac{7\pi}{6} = -\frac{1}{2}$$

$$\cos\frac{7\pi}{6} = -\frac{\sqrt{3}}{2}$$

Use a definition of the tangent to find its value:

$$\tan\frac{7\pi}{6} = \frac{\sin\dfrac{7\pi}{6}}{\cos\dfrac{7\pi}{6}} = \frac{-\dfrac{1}{2}}{-\dfrac{\sqrt{3}}{2}} = \frac{1\cdot2}{\sqrt{3}\cdot2} = \frac{1\cdot\sqrt{3}}{\sqrt{3}\cdot\sqrt{3}} = \frac{\sqrt{3}}{3}$$

Step 3: Find the three reciprocal functions by taking reciprocals of the first three:

$$\csc\frac{7\pi}{6} = \frac{1}{\sin\dfrac{7\pi}{6}} = \frac{1}{-\dfrac{1}{2}} = -2$$

$$\sec\frac{7\pi}{6} = \frac{1}{\cos\dfrac{7\pi}{6}} = \frac{1}{-\dfrac{\sqrt{3}}{2}} = -\frac{2}{\sqrt{3}} = -\frac{2\sqrt{3}}{3}$$

$$\cot\frac{7\pi}{6} = \frac{1}{\tan\dfrac{7\pi}{6}} = \frac{1}{\dfrac{\sqrt{3}}{3}} = \frac{3}{\sqrt{3}} = \frac{3\sqrt{3}}{3} = \sqrt{3}$$

Solution: The trigonometric functions are: $\sin\dfrac{7\pi}{6} = -\dfrac{1}{2}$, $\cos\dfrac{7\pi}{6} = -\dfrac{\sqrt{3}}{2}$, $\tan\dfrac{7\pi}{6} = \dfrac{\sqrt{3}}{3}$, $\csc\dfrac{7\pi}{6} = -2$, $\sec\dfrac{7\pi}{6} = -\dfrac{2\sqrt{3}}{3}$, $\cot\dfrac{7\pi}{6} = \sqrt{3}$.

Sometimes, you will be asked to find the trigonometric functions for angles that are not placed on the unit circle. Often, the requested angles are the coterminal angles of the ones that are located on the unit circle. Your task then is to identify the corresponding coterminal angles and calculate their trigonometric functions. Let's see how it works with the next problems.

Sample Problem 2

Find the six trigonometric functions for $\dfrac{9\pi}{4}$.

Step 1: Because this angle is not placed on the unit circle, you should suspect that its measure is greater than 360° or 2π. You can rewrite this angle as follows:

$$\frac{9\pi}{4} = \frac{8\pi}{4} + \frac{\pi}{4} = 2\pi + \frac{\pi}{4}$$

You find that the original angle $\dfrac{9\pi}{4}$ is coterminal with $\dfrac{\pi}{4}$; therefore, its trigonometric functions are the same as for $\dfrac{\pi}{4}$.

Step 2: Find the three main trigonometric functions for $\dfrac{\pi}{4}$ using the coordinates on the unit circle $\left(\dfrac{\sqrt{2}}{2}, \dfrac{\sqrt{2}}{2} \right)$:

$$\sin \frac{\pi}{4} = \frac{\sqrt{2}}{2}$$

$$\cos \frac{\pi}{4} = \frac{\sqrt{2}}{2}$$

$$\tan \frac{\pi}{4} = \frac{\dfrac{\sqrt{2}}{2}}{\dfrac{\sqrt{2}}{2}} = 1$$

Step 3: Find the three other reciprocal functions:

$$\csc \frac{\pi}{4} = \frac{1}{\sin \dfrac{\pi}{4}} = \frac{1}{\dfrac{\sqrt{2}}{2}} = \frac{2}{\sqrt{2}} = \frac{2 \cdot \sqrt{2}}{\sqrt{2} \cdot \sqrt{2}} = \sqrt{2}$$

$$\sec \frac{\pi}{4} = \frac{1}{\cos \dfrac{\pi}{4}} = \frac{1}{\dfrac{\sqrt{2}}{2}} = \frac{2}{\sqrt{2}} = \frac{2 \cdot \sqrt{2}}{\sqrt{2} \cdot \sqrt{2}} = \sqrt{2}$$

$$\cot \frac{\pi}{4} = \frac{1}{\tan \dfrac{\pi}{4}} = \frac{1}{1} = 1$$

Solution: The trigonometric functions are: $\sin \dfrac{9\pi}{4} = \dfrac{\sqrt{2}}{2}$, $\cos \dfrac{9\pi}{4} = \dfrac{\sqrt{2}}{2}$, $\tan \dfrac{9\pi}{4} = 1$, $\csc \dfrac{9\pi}{4} = \sqrt{2}$, $\sec \dfrac{9\pi}{4} = \sqrt{2}$, $\cot \dfrac{9\pi}{4} = 1$.

Sample Problem 3

Find the tangent of 3,720°.

Step 1: A full revolution is only 360°, so in this case, there is definitely more than one revolution. You can use the method identified before while discussing coterminal angles and start subtracting 360° until your angle is less than 360°. But that would be a lengthy process. Are there any shortcuts? Certainly there are. Instead of subtraction, you can use division and multiplication. Let's divide 3,720 by 360 to see how many times 360 goes into 3,720:

$$3,720 \div 360 = 10.333\ldots$$

It cannot be divided evenly, but you already have what you need. You can see that 360 goes into 3,720 10 times with the remainder that you are not interested in. That means that there were 10 full revolutions plus some additional rotation that was less than 360°.

Step 2: Multiply 360 by 10 times to see how many degrees these 10 revolutions are:

$$360 \cdot 10 = 3,600$$

Subtract 3,600 from 3,720 to find the angle that is less than 360°:

$$3,720 - 3,600 = 120$$

The angle 3,720° is coterminal with angle 120°. This means that they have the same values for trigonometric functions. Now you need to find the tangent of 120° to answer the problem's question.

Step 3: From the unit circle, you can get the values for the sine and cosine of 120°. They are $\left(-\dfrac{1}{2}, \dfrac{\sqrt{3}}{2}\right)$. Use the definition of the tangent and substitute the values:

$$\tan 120° = \frac{\sin 120°}{\cos 120°} = \frac{\dfrac{\sqrt{3}}{2}}{-\dfrac{1}{2}} = -\frac{\sqrt{3} \cdot 2}{1 \cdot 2} = -\sqrt{3}$$

The tangent of $120° = -\sqrt{3}$.

DANGEROUS TURN

When using the coordinates from the unit circle to find the values of trigonometric functions, keep in mind that the first coordinate is always the cosine and the second coordinate is the sine. If you use them in the wrong order, the answer will be incorrect.

Step 4: Always check your answer by recalling the signs of the trigonometric functions in different quadrants. 120° is in the second quadrant, and as you remember, the tangent is always negative for the angles in the second quadrant. So the answer is correct in terms of the sign of the function. The last step is to state that because two angles 120° and 3,720° are coterminal, they have the same values of trigonometric functions. So the tangent of 3,720° $= -\sqrt{3}$.

Solution: The tangent of 3,720° $= -\sqrt{3}$.

Let's do another problem that uses the same approach of working with trigonometric functions.

Sample Problem 4

Identify whether the sine of 100,000° is positive or negative.

Step 1: Subtracting 360° will take too long, so resort to the previous method. Divide 100,000 by 360:

$$100,000 \div 360 = 277.777\ldots$$

Step 2: Multiply 360° by 277 full revolutions and subtract the resulting product from 100,000 to get the angle that is less than 360°:

$$360 \cdot 277 = 99,720$$

$$100,000 - 99,720 = 280$$

The coterminal angle that is less than 360° is equal to 280°.

Step 3: This means that the sign of the sine function for 100,000° is the same as for 280°. Angle 280° belongs to the fourth quadrant; therefore, the sine of this angle is negative. Thus, the sine of 100,000° is negative as well.

Solution: The sine of 100,000° is negative.

As this chapter has illustrated, the unit circle perspective of trigonometry is one that complements the triangle-based approach and covers cases such as negative or very large angles that cannot be easily illustrated on a triangle. Utilizing the unit circle also helps remember useful shortcuts like ones for determining the sign of a trigonometric function or realizing that sines and cosines of several common angles have 0 and 1 values.

Practice Problems

Problem 1: Show that the coordinates for the angle of $\dfrac{5\pi}{4}$ satisfy the unit circle's equation.

Problem 2: Find the cotangent of 1,845°.

Problem 3: Determine whether the cosine of 25,000° is positive or negative.

Problem 4: Find six trigonometric functions for $\dfrac{4\pi}{3}$.

Problem 5: Find six trigonometric functions for $\dfrac{5\pi}{2}$.

The Least You Need to Know

- Using the second historical approach, you can visualize and calculate the trigonometric functions of all angles with values from negative infinity to positive infinity.

- The sine is positive in the first and second quadrants and negative in the third and fourth quadrants.

- The cosine is positive in the first and fourth quadrants and negative in the second and third quadrants.

- Some of the commonly used values for trig functions are sin0° = 0, cos0° = 1, sin90° = 1, and cos90° = 0.

Trigonometric Functions of Any Angle

Chapter 10

In This Chapter

- Finding the reference angles
- Evaluating trigonometric functions for any angle
- Prefixing the appropriate sign to the function value
- Distinguishing between even and odd trig functions

In Chapter 9, you learned how to find the trigonometric function of angles that are either coterminal with (or multiples of) the commonly used special angles $\frac{\pi}{6}$ (30°), $\frac{\pi}{4}$ (45°), and $\frac{\pi}{3}$ (60°). In this chapter, we expand the repertoire of angles for which you can find trigonometric functions by introducing reference angles. We also explore the notion of even and odd functions.

Calculating Values of Trigonometric Functions

People who study trigonometry in modern times can be considered lucky because they have powerful resources to find values for trigonometric functions of all angles. All you have to do is to punch several buttons on your graphing calculator and—voilà!—the values of any trigonometric function for any angle are displayed in a fraction of a second. You can read about the use of a graphing calculator for this purpose in Chapter 22.

What if you do not have a graphing calculator handy—or the battery dies? In modern times, you are still safe if you have access to the Internet. You can go, for example, to Google, type in the trig function of any angle, and the answer will be available immediately as well.

One of the authors remembers her experience studying trigonometry in ninth grade with no graphing calculator or Internet (neither came into widespread use until the early '90s). Students had to rely on the trigonometric tables that every trigonometry textbook had in the back. And finding the corresponding values of the trigonometric functions was not that easy—you had to follow some rules and had to be careful to look at the correct row and column in the table.

Most modern textbooks still include the trigonometric tables. But why do you need the trigonometric tables now if you have such powerful tools, like graphing calculators and the Internet, at your disposal? The answer to this question is similar to the answer to the question of why it's important to learn the multiplication tables or long division.

The narrow answer to all these questions (including the values of trig functions) is that your teachers often will not allow you to use graphing calculators or the Internet while taking quizzes or tests. The broader answer is more philosophical—you need to understand how mathematics works, not just how to punch some buttons to get answers.

Let's go back to the question at hand about the use of the trigonometric tables as a resource for finding the values of trigonometric functions. If you already looked in the back of this book, you might have noticed that the trigonometric tables have the values for angles ranging from 0° to 90°. What about all other angles? Remember, we promised you would learn to find the values of trigonometric functions for all angles.

This is where the *reference angles* come in. The next section discusses them.

DEFINITION

The acute angle θ is a **reference angle** for the angles $180° - \theta$, $180° + \theta$, $360° - \theta$, and all coterminal angles.

Reference Angles

Let θ be an acute angle in standard position. Suppose that $\theta = 40°$. Notice that on the following figure the terminal ray of $\theta = 40°$ and the terminal ray of $180° - \theta = 140°$ are symmetrical in the y-axis.

If the sine and cosine of θ are known, then the sine and cosine of 140° can be deduced. The angle $\theta = 40°$ is the reference angle for 140°.

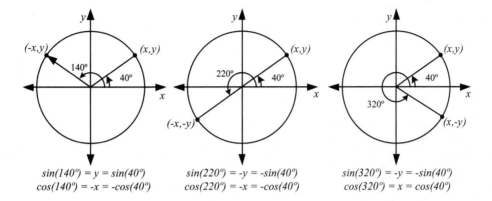

$$sin(140°) = y = sin(40°)$$
$$cos(140°) = -x = -cos(40°)$$

$$sin(220°) = -y = -sin(40°)$$
$$cos(220°) = -x = -cos(40°)$$

$$sin(320°) = -y = -sin(40°)$$
$$cos(320°) = x = cos(40°)$$

It is also the reference angle for the 220° and 320° angles shown in the middle and in the right-most drawings.

You can then say that the reference angle for any angle θ is the acute positive angle θ' formed by the terminal ray of θ and the x-axis.

The following table summarizes the procedure of finding the reference angle for angles in different quadrants.

The first column displays a quadrant of the angle for which you need to find the trigonometric functions. The second column shows the measure of the angle of interest in both degrees and in radians. The third column identifies the formula by which the reference angle can be found. The fourth, fifth, and sixth columns show the signs for the sine, cosine, and tangent for the final answer.

Quadrant	Measure of θ	Reference angle	Sine	Cosine	Tangent
I	0° to 90 0 to $\dfrac{\pi}{2}$	θ	+	+	+
II	90° to 180° $\dfrac{\pi}{2}$ to π	$180° - \theta$ $\pi - \theta$	+	–	–
III	180° to 270° π to $\dfrac{3\pi}{2}$	$\theta - 180°$ $\theta - \pi$	–	–	+

Quadrant	Measure of θ	Reference angle	Sine	Cosine	Tangent
IV	270° to 360°	360° − θ			
	$\dfrac{3\pi}{2}$ to 2π	$2\pi - \theta$	−	+	−

We advise that you study this table and, with time, memorize some of the steps you need to find the values for trigonometric functions of any angle.

WORTH KNOWING

The signs of the reciprocal functions in each quadrant are the same as the signs of the corresponding main trigonometric functions.

Before you solve sample problems, let's draw your attention to the second-quadrant angles in this table. Do you recall that when studying the trigonometric functions in Chapter 6 by using right triangles, we produced two statements that we called "the ambiguous case"? One was that the sine of the obtuse angle is the same as the sine of that angle's supplement. The second statement was that the cosine of the obtuse angle is equal to the cosine of that angle's supplement multiplied by –1.

You get the same results with the current approach. The third column provides the formula to find these supplements, and the fourth and fifth columns provide the correct signs for the final answer. Certainly, there are no surprises here—both approaches are supposed to work perfectly by giving the same values for trigonometric functions of a particular angle regardless of the way you calculate them.

Evaluating Trigonometric Functions

Equipped with the knowledge on reference angles, you can apply it now to find trigonometric functions of different angles.

When asked to find trigonometric functions for an angle, you have to run some simple tests to identify the best strategy to find the answer the quickest way.

First, make sure that the angle of interest is not one of the special angles you've seen so many times. If it is, then you should know the answer right away.

The following problems illustrate how you should handle all other angles.

Sample Problem 1

Find the sine, cosine, and tangent for an angle of 225°.

Step 1: Identify the quadrant. This angle is between 180° and 270°, so the terminal side is in the third quadrant.

Step 2: Perform the operation indicated for that quadrant in the table to find the reference angle. You need to subtract 180° from the angle:

$$\theta - 180°$$

$$225° - 180° = 45°$$

It turns out that the reference angle is one of the special angles for which you know the values of trigonometric functions.

Step 3: Write down the value of three trigonometric functions for the reference angle:

$$\sin 45° = \frac{\sqrt{2}}{2}$$

$$\cos 45° = \frac{\sqrt{2}}{2}$$

$$\tan 45° = 1$$

Step 4: As the final step, look for the signs of trigonometric functions for angles in the third quadrant. Both the sine and cosine are negative for the angles in the third quadrant. That means that the tangent will be positive (as the table states) because you need to divide two negative values to obtain it:

$$\cos 225° = -\frac{\sqrt{2}}{2}$$

$$\sin 225° = -\frac{\sqrt{2}}{2}$$

$$\tan 225° = 1$$

Solution: The trigonometric functions values are thus $\sin 225° = -\dfrac{\sqrt{2}}{2}$, $\cos 225° = -\dfrac{\sqrt{2}}{2}$, and $\tan 225° = 1$.

In the next problem, an angle is given in radians. Let's see how it works.

Sample Problem 2

Find the sine, cosine, and tangent for an angle measuring $\dfrac{29\pi}{3}$.

Step 1: This angle is certainly more than 2π (360°) because $29\pi \div 3 > 2\pi$. This means that you first need to find a coterminal angle that is less than 360°.

WORTH KNOWING

One anonymous teacher developed this little trick as an aid to remember the signs for trigonometric functions in different quadrants. Each quadrant is assigned a letter: I has A, II has S, III has T, and IV has C. To remember the letters, you can create any silly sentence—for example, "All sports train character." This letter will identify the first letter of the name of the function that is positive in this quadrant. Note that the reciprocal of the function has the same sign.

A: All functions are positive.

S: Sine and its reciprocal, the cosecant, are positive.

T: Tangent and its reciprocal, the cotangent, are positive.

C: Cosine and its reciprocal, the cosecant, are positive.

One full revolution is 2π, and four full revolutions are 8π. 8π is the best choice because it is the next smallest even multiple of π after 10π, which is too large for us since, multiplied by 3, it would equal to more than 29π. Only even multiples are in consideration so that the angle as a whole can be evenly divided by 2π. Therefore, you can rewrite the angle as:

$$\frac{29\pi}{3} = \frac{24\pi}{3} + \frac{5\pi}{3} = 8\pi + \frac{5\pi}{3} = 4(2\pi) + \frac{5\pi}{3}$$

The coterminal angle is $\dfrac{5\pi}{3}$.

Step 2: Identify the quadrant. This angle is between $\dfrac{3\pi}{2}$ and 2π, so the terminal side is in the fourth quadrant.

Step 3: To find the reference angle, perform the operation indicated by the table for that quadrant. You need to subtract the angle from 2π to find the reference angle:

$$2\pi - \theta$$
$$2\pi - \frac{5\pi}{3} = \frac{6\pi}{3} - \frac{5\pi}{3} = \frac{\pi}{3}$$

It again turns out that the reference angle is one of the special angles for which you know the values of trigonometric functions.

Step 4: Write down the values of the three trigonometric functions for the reference angle:

$$\sin\frac{\pi}{3} = \frac{\sqrt{3}}{2}$$

$$\cos\frac{\pi}{3} = \frac{1}{2}$$

$$\tan\frac{\pi}{3} = \sqrt{3}$$

Step 5: As the final step, look for the signs of trigonometric functions for angles in the fourth quadrant. The sine is negative and the cosine is positive for the angles in the fourth quadrant. That means that the tangent will be negative because you need to divide a negative number by a positive:

$$\sin\frac{5\pi}{3} = -\frac{\sqrt{3}}{2}$$

$$\cos\frac{5\pi}{3} = \frac{1}{2}$$

$$\tan\frac{5\pi}{3} = -\sqrt{3}$$

Solution: The trigonometric function values are thus $\sin\frac{5\pi}{3} = -\frac{\sqrt{3}}{2}$, $\cos\frac{5\pi}{3} = \frac{1}{2}$, and $\tan\frac{5\pi}{3} = -\sqrt{3}$.

Let's do two more problems on finding reference angles.

Sample Problem 3

Find the sine, cosine, and tangent for an angle of 163°.

Step 1: Identify the quadrant. This angle is between 90° and 180°, so the terminal side is in the second quadrant.

DANGEROUS TURNS

Don't try to determine the trigonometric functions for an angle that is not in standard position. You will not be able to identify a correct reference angle for it; therefore, the values of the trigonometric functions will be wrong.

Step 2: Perform the operation indicated for that quadrant in the table to find the reference angle. You need to subtract the angle from 180°:

$$180° - \theta$$

$$180° - 163° = 17°$$

It turns out that the reference angle is not one of the special angles, so you need to look up its trigonometric functions in the trigonometric tables in Appendix D of this book.

Step 3: Write down the values of three trigonometric functions for the reference angle:

$$\sin 17° = 0.2924$$

$$\cos 17° = 0.9563$$

$$\tan 17° = 0.3057$$

Step 4: As the final step, look for the signs of trigonometric functions for angles in the second quadrant. The sine is positive and the cosine is negative for the angles in the second quadrant. That means that the tangent will be negative because you need to divide a positive number by a negative:

$$\sin 163° = 0.2924$$

$$\cos 163° = -0.9563$$

$$\tan 163° = -0.3057$$

Solution: The trigonometric function values are thus sin163° = 0.2924, cos163° = –0.9563, and tan163° = –0.3057.

For the last problem in this section, we chose a negative angle.

Sample Problem 4

Find the cosecant of –210°.

Step 1: Because –210° + 360° = 150°, it follows that –210° is coterminal with the angle of 150°.

Step 2: The angle of 150° is a second-quadrant angle, so you can find the reference angle as follows:

$$180° - 150° = 30°$$

The reference angle is 30°.

Step 3: 30° is one of the special angles and its sine is known to you:

$$\sin 30° = \frac{1}{2}$$

Because the angle of 150° is in the second quadrant, its sine is positive—as is the sign of its reciprocal, the cosecant:

$$\sin 150° = \frac{1}{2}$$
$$\csc 150° = 2$$

Because 150° is coterminal with the original angle of –210°, you can state that the cosecant of –210° is 2.

Solution: The cosecant of –210° is 2.

The last point that we want to make is about reciprocal functions. The sample problems mostly focus on three main trigonometric functions—the sine, the cosine, and the tangent. The other three functions—the cosecant, the secant, and the cotangent—can be found easily by just taking the reciprocals. For example, if in sample problem 1 you also need to find the value of the secant, you just need to take the reciprocal of the cosine value:

$$\cos 225° = -\frac{\sqrt{2}}{2}$$

$$\sec 225° = \frac{1}{\sin 225°} = \frac{1}{-\dfrac{\sqrt{2}}{2}} = -\frac{2}{\sqrt{2}} = -\frac{2\sqrt{2}}{\sqrt{2}\cdot\sqrt{2}} = -\frac{2\sqrt{2}}{2} = -\sqrt{2}$$

Keep in mind that reciprocal functions have the same sign in each quadrant as their corresponding main trigonometric functions.

Looking Closely at Trigonometric Functions

After discussing the values of trigonometric functions for so many angles, it is time to look more closely at these functions and make some generalizations about them.

Periodic Functions

You already know that if the sine of some angle is equal to 1, then this angle is 90°, because sin90° = 1. But because any angle coterminal with 90° also has 1 as its sine value, you can state that all angles such as

90°, 90° + 360°, 90° + 2 · 360°, 90° + 3 · 360°, …

are solutions to the following equation:

$\sin\theta = 1$

These angles can be written more conveniently as $\theta = 90° + n \cdot 360°$, where n is an integer.

In radians, the same statement can be written as follows:

$$\theta = \frac{\pi}{2} + 2n\pi$$

The same is true for the cosine function. Both the sine and cosine functions repeat their values every 360° or 2π radians. These facts can be summarized by saying that the sine and cosine functions are *periodic functions* and that they have a *fundamental period* of 360° or 2π radians.

> **DEFINITION**
>
> A function f is a **periodic function** if there is a positive number p, called a period of f, such as that $f(x + p) = f(x)$ for all x in the domain of x. The smallest period of a periodic function is called the **fundamental period** of the function.

We explore this further in Chapters 11 and 12. Now let's look at another feature of the trigonometric functions.

Even and Odd Trigonometric Functions

Let's explore the relationship between $\sin\theta$ and $\sin(-\theta)$. Let θ be 30°; then, $-\theta = -30°$. Using the procedures discussed in the previous sample problems, you can show that:

$$\sin(-\theta) = -\sin(\theta)$$

You know that $\sin 30° = \dfrac{1}{2}$ and that $-30°$ is coterminal with:

$$-30° + 360° = 330°$$

Angle 330° is a fourth-quadrant angle, so its reference angle can be found by:

$$360° - 330° = 30°$$

So the reference angle for the original angle of $-30°$ is a 30° angle, and the sign in the fourth quadrant is negative, so you can write:

$$\sin(-30°) = -\dfrac{1}{2}$$

You proved that:

$$\sin(-30°) = -\sin 30°$$
$$-\dfrac{1}{2} = -\dfrac{1}{2}$$

The statement $\sin(-\theta) = -\sin(\theta)$ is true for any angle.

However, the cosine function is different because it is an even function. You have:

$$\cos(-\theta) = \cos(\theta)$$

For example, $\cos(-60°) = \dfrac{1}{2}$, which is equal to $\cos(60°) = \dfrac{1}{2}$.

In general, you can distinguish two types of functions: *even functions* and *odd functions*.

> **DEFINITION**
>
> A function is called **even** if, for every x, $f(-x) = f(x)$.
>
> A function is called **odd** if, for every x, $f(-x) = -f(x)$.

The tangent function is the sine divided by the cosine—an odd function divided by an even function—therefore, it is odd. The cosecant is the reciprocal of an odd function (the sine), so it is naturally also an odd function.

In summary, you can say that the cosine and the secant are even functions, whereas the sine, the cosecant, the tangent, and the cotangent are odd functions. Therefore, there are two even and four odd trigonometric functions. In Chapter 11, we see how the graphs of trigonometric functions reflect the fact of being either an even or an odd function.

Practice Problems

Problem 1: Find the sine, cosine, and tangent for an angle of 834°.

Problem 2: Find the sine, cosine, and tangent for an angle of $\dfrac{34\pi}{3}$.

Problem 3: Find the sine, cosine, and tangent for an angle of 150°.

Problem 4: Find the cosecant of −132°.

Problem 5: Find the cotangent of −405°.

The Least You Need to Know

- The introduction of a reference angle enables you to find values for trig functions of any angle.
- All functions are positive in the first quadrant.
- The sine and the cosine are positive in the first quadrant; the sine and cosecant are positive in the second quadrant; the tangent and cotangent are positive in the third quadrant; and the cosine and secant are positive in the fourth quadrant.
- The cosine and the secant are even functions, whereas the sine, the cosecant, the tangent, and the cotangent are odd functions.

Graphs of Trigonometric Functions

This part presents a gallery of "portraits" or graphs of all six trigonometric functions, their variations, and their inverses.

First, you learn how to sketch basic trigonometric functions. Then, you explore how to stretch and shrink the graphs along the x- and y-axes, as well as how to shift the graphs along both axes. You learn that adding a number to the function or multiplying the function by a number is quite different from adding a number to the angle or multiplying the angle of the function by a number.

Later, you look at the graphs of six inverse functions and how to find their exact values. Finally, you work with the compositions of trigonometric and inverse trigonometric functions such as sin(arcsin$x$) and learn that dealing with them and finding their values is easier than it first seems.

Graphs of Sine and Cosine Functions

In This Chapter

- Exploring the basic sine and cosine curves
- Understanding how shifting and stretching of graphs works
- Translating the graphs along the *x*- and *y*-axes
- Using the sine and cosine curves to model real-life situations

You already know a lot about the sine and cosine functions. You know their reciprocals and how to determine the values of the sine and cosine for any angle. You also learned that both functions are periodic functions, and that the sine is an odd function and the cosine is an even function. But something is still missing from the big picture. From algebra, you probably remember that functions were discussed with graphs. We have not discussed graphs yet, but that's about to change. In this chapter, we discuss and explore the graphs of the sine and cosine functions.

Basic Sine Curve

This section presents techniques for sketching a graph of the sine function, which is usually called a sine curve. Two axes on the coordinate plane represent two different kinds of values. The *x*-axis is labeled in either degrees or radians, and the *y*-axis is labeled in real numbers.

Domain and Range of the Sine Curve

Before we sketch the sine curve, as with all functions, let's talk about this function's domain and range. Because you can identify the sine function for any angle, the domain, or x-values, of the sine function include all angles in degrees or radians (both of which span the range of real numbers), so the sine curve rolls smoothly without any interruption and repeats itself as a periodic function.

To determine the range or the y-values of the sine function, recall that the unit circle has a radius of 1, or $r = 1$; and by definition, $\sin \theta = y$. Therefore, the values for y range between –1 and 1 are:

$$-1 \le y \le 1$$

$$-1 \le \sin \theta \le 1$$

Key Points on Basic Sine Curve

To construct the graph of the sine curve, it helps to imagine a grain of sand on the unit circle that starts at (1,0) and rotates counterclockwise around the origin. Every position (x,y) of the grain sand corresponds to an angle θ, where $\sin \theta = y$.

As the grain of sand rotates through the four quadrants, shown on the following figure, you obtain the four pieces of the following sine graph:

From 0 to $\dfrac{\pi}{2}$, the y-coordinate increases from 0 to 1.

From $\dfrac{\pi}{2}$ to π, the y-coordinate decreases from 1 to 0.

From π to $\dfrac{3\pi}{2}$, the y-coordinate decreases from 0 to –1.

From $\dfrac{3\pi}{2}$ to 2π, the y-coordinate increases from –1 to 0.

There are five key points (x,y) in one period of the sine graph where the sine graph has its maximum ($\sin \theta = 1$), minimum ($\sin \theta = -1$), and where it intercepts (crosses) the x-axis ($\sin \theta = 0$). The x-coordinates of these key points correspond to the quadrant angles as shown in the following table.

Intercept	Maximum	Intercept	Minimum	Intercept
(0,0)	$\left(\dfrac{\pi}{2},1\right)$	$(\pi,0)$	$\left(\dfrac{3\pi}{2},-1\right)$	$(2\pi,0)$
(0,0)	(90°,1)	(180°,0)	(270°,–1)	(360°,0)

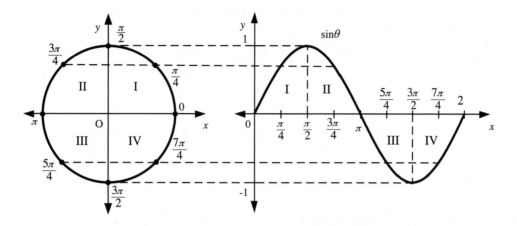

Because the sine function is a periodic function with the fundamental period of 2π (360°), the graph can be extended left and right along the x-axis as illustrated in the following.

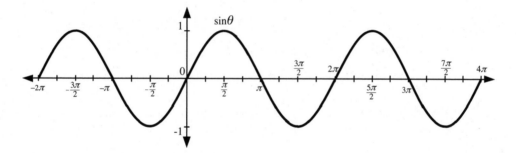

The graph continues indefinitely in both directions, repeating itself over and over again.

Other Sinusoidal Curves

One of the most important uses of trigonometry is to describe periodic processes such as the tidal movement of the ocean, changes of body temperature, the length of the day throughout the year, the swing of a pendulum, and many more.

All of these periodic motions can be described by a family of functions of the form:

$$y = a\sin(bx - c) + d$$

where a, b, c, and d are constants.

The graphs of these functions are called *sinusoidal curves*. They are closely related to the simple sine curve that we've already discussed but they go through some shifting and stretching. Let's discuss everything step by step, analyzing each of the constants *a*, *b*, *c*, and *d*.

DEFINITION

Sinusoidal curves are graphs of the sine or cosine in the general form of $y = a\sin(bx - c) + d$ and $y = a\cos(bx - c) + d$.

When *a* and *b* are both equal to 1 and *c* and *d* are both 0, you obtain the simple sine curve $y = \sin x$.

The Amplitude and Vertical Stretching and Shrinking

In the equation $y = a\sin(bx - c) + d$, the constant *a* is called the *amplitude* of the function. It tells you how far from zero the values of the function along the *y*-axis can go.

DEFINITION

The **amplitude** of $y = a\sin(bx - c) + d$ is the largest value of *y* and is given by: amplitude = |a|.

When |a|, or the absolute value of *a*, is more than 1, the basic sine curve is stretched; when |a| is less than 1, it is shrunk.

Sample Problem 1

Sketch the graph of $y = 3\sin x$.

Step 1: The values of this function are three times the corresponding values of the basic sine function $y = \sin x$. Thus, each *y*-value is multiplied by 3.

Step 2: Sketch the graph by placing five key points. Keep in mind that the maximum becomes 3 and the minimum becomes –3, but the intercept points stay the same because $0 \cdot 3 = 0$.

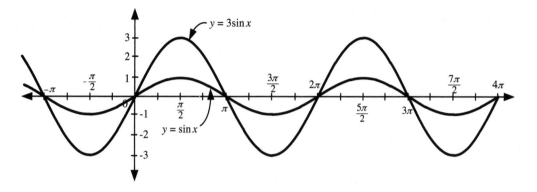

Solution: The graph of $y = 3\sin x$ is obtained from the basic sine curve by vertically stretching it in the y-direction.

Sample Problem 2

Sketch the graph of $y = \dfrac{1}{2}\sin x$.

Step 1: The values of this function are two times less than the corresponding values of the basic sine function $y = \sin x$ because each y-value is multiplied by $\dfrac{1}{2}$.

Step 2: Sketch the graph by placing five key points. Keep in mind that the maximum becomes $\dfrac{1}{2}$ $(1 \cdot \dfrac{1}{2} = \dfrac{1}{2})$ and the minimum becomes $-\dfrac{1}{2}$, but the intercept points stay the same because $0 \cdot \dfrac{1}{2} = 0$:

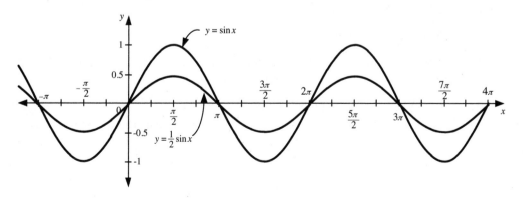

Solution: The graph of $y = \dfrac{1}{2}\sin x$ is obtained from the basic sine curve by vertically shrinking it in the y-direction.

Horizontal Stretching and Shrinking

In this section, we explore the effect of the constant b, which is called the *frequency* of the curve. It is responsible for horizontal stretching and shrinking.

DEFINITION

Let **frequency** b be a positive real number. It tells you how many periods are repeated in an interval of 2π. The period can be found by dividing 2π by b.

The period of $y = \sin bx$ is $\dfrac{2\pi}{b}$.

If $0 < b < 1$, then the period of $y = \sin bx$ is greater than 2π and the graph is stretched horizontally.

If $b > 1$, then the period of $y = \sin bx$ is less than 2π and the graph is shrunk horizontally.

Sample Problem 3

Sketch the graph of $y = \sin \dfrac{x}{2}$.

Step 1: The amplitude is 1 and the period can be found as:

$$2\pi \div \frac{1}{2} = \frac{2\pi}{1} \div \frac{1}{2} = \frac{2\pi}{1} \cdot \frac{2}{1} = 4\pi$$

This means that the graph is stretched and the key points have different coordinates.

Step 2: Find the five key points. When b is between 0 and 1, the period of the function will always be more than 2π. In this case it is 4π, or two times more than 2π. This means that the x-coordinate of each key point should be multiplied by 2. Then, the key points become as follows:

$$(0,0),\ (\pi,1),\ (2\pi,0),\ (3\pi,-1),\ (4\pi,0)$$

Step 3: Sketch the graph using key points.

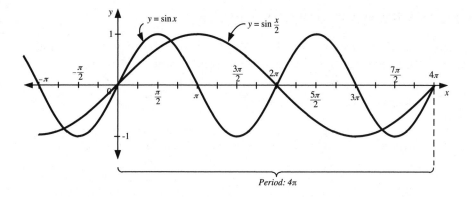

Period: 4π

Solution: The graph of $y = \sin\dfrac{x}{2}$ is obtained from the basic sine curve by horizontally stretching it in the x-direction.

Sample Problem 4

Sketch the graph of $y = \sin 3x$.

Step 1: The amplitude is 1 and, because b is 3, the period can be found as:

$$2\pi \div 3 = \frac{2\pi}{3}$$

This means that the graph is shrunk because the whole cycle should now fit into $\dfrac{2\pi}{3}$ instead of 2π. Therefore, the key points have different coordinates.

Step 2: Find the five key points. When b is a whole number, the period of the function is always less than 2π. In this case, it is $\dfrac{2\pi}{3}$ or three times less than 2π. This means that the x-coordinate of each key point of the basic sine curve should be divided by 3. Then, the key points become as follows:

$$(0,0), \ \left(\frac{\pi}{6},1\right), \ \left(\frac{\pi}{3},0\right), \ \left(\frac{\pi}{2},-1\right), \ \left(\frac{2\pi}{3},0\right)$$

Let's show how you got the new x-coordinate of the fourth key point. For the basic sine curve, this point was $\left(\dfrac{3\pi}{2},-1\right)$. Divide the x-coordinate by 3:

$$\frac{3\pi}{2} \div 3 = \frac{3\pi}{2} \div \frac{3}{1} = \frac{3\pi}{2} \cdot \frac{1}{3} = \frac{\pi}{2}$$

Note that the y-coordinate stays the same since the transformation of the graph is completely horizontal. All other key points were found similarly.

Step 3: Sketch the graph using new key points.

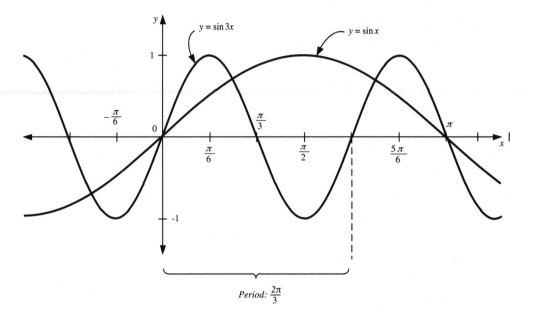

Period: $\frac{2\pi}{3}$

Solution: The graph of $y = \sin3x$ is obtained from the basic sine curve by horizontally shrinking it in the x-direction.

Summarizing the discussion in the previous two sections, we would like to emphasize one more time that the two constants, a (amplitude) and b (frequency), work differently. When a (amplitude) is a whole number, the curve stretches vertically, but when b (frequency) is a whole number, the curve shrinks horizontally. And vice versa, when a is a value between 0 and 1 and 0 and –1, the curve shrinks vertically, but when b is a value between 0 and 1, the curve stretches horizontally.

Translations of the Sine Curve

In addition to the two cases of altering the basic sine curve by vertical and horizontal stretching and shrinking, you have two more possibilities. You can also shift the curve up and down along the y-axis or shift it left or right along the x-axis. These shifts are called *translations* of the curve.

DEFINITION

Translation of a function involves a combination of shifts of its graph up or down the y-axis and shifts left or right along the x-axis. Note that one of the shifts can be by 0 units, which, in effect, translates the function only in one direction.

Horizontal Shift

Let's look again at the general equation of the sinusoidal curves:

$$y = a\sin(bx - c) + d$$

Constant c is called the *phase* or *phase shift* of the curve.

> **DEFINITION**
>
> The **phase** of the curve tells you how much the curve has been shifted left or right (horizontal shift). The number $\dfrac{c}{b}$ is called the **phase shift.** The left and right endpoints corresponding to a one-cycle interval of the graph can be calculated by solving the equations: $bx - c = 0$ and $bx - c = 2\pi$.

Sample Problem 5

Sketch the graph of $y = \dfrac{1}{2}\sin\left(x - \dfrac{\pi}{3}\right)$.

Step 1: The amplitude is $\dfrac{1}{2}$ and, because b is 1, the period is a regular period of 2π. Because of the amplitude, the y-coordinate of each key point is two times less.

Step 2: Find the x-coordinates of five key points. Solve the equations discussed in the definition of the phase shift to find the end points (note that $b = 1$ and $c = \dfrac{\pi}{3}$):

$$bx - c = 0 \rightarrow x - \frac{\pi}{3} = 0 \text{, thus } x = \frac{\pi}{3}$$

$$bx - c = 2\pi \rightarrow x - \frac{\pi}{3} = 2\pi \rightarrow x = 2\pi + \frac{\pi}{3} = \frac{6\pi}{3} + \frac{\pi}{3} = \frac{7\pi}{3} \text{, thus } x = \frac{7\pi}{3}$$

The first cycle of the curve starts at $x = \dfrac{\pi}{3}$ and ends at $x = \dfrac{7\pi}{3}$. Recall that the basic curve starts at $x = 0$ and ends at $x = 2\pi$.

Step 3: To find the other three x-coordinates of key points, you need to add the phase shift to each x-coordinate. Because b is 1, the phase shift is $\dfrac{c}{b} = \dfrac{c}{1} = c = \dfrac{\pi}{3}$.

Second key point: $\dfrac{\pi}{2} + \dfrac{\pi}{3} = \dfrac{3\pi}{6} + \dfrac{2\pi}{6} = \dfrac{5\pi}{6}$

Third key point: $\pi + \dfrac{\pi}{3} = \dfrac{3\pi}{3} + \dfrac{\pi}{3} = \dfrac{4\pi}{3}$

Fourth key point: $\dfrac{3\pi}{2} + \dfrac{\pi}{3} = \dfrac{9\pi}{6} + \dfrac{2\pi}{6} = \dfrac{11\pi}{6}$

Note that all five key y-coordinates are divided by 2 because the amplitude is $\frac{1}{2}$. Then, the key points become as follows:

$$\left(\frac{\pi}{3},0\right), \left(\frac{5\pi}{6},\frac{1}{2}\right), \left(\frac{4\pi}{3},0\right), \left(\frac{11\pi}{6},-\frac{1}{2}\right), \left(\frac{7\pi}{3},0\right)$$

Step 4: Sketch the graph using the new key points.

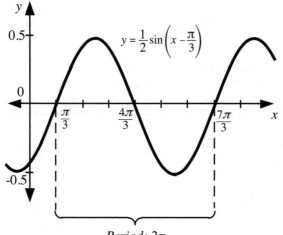

$$y = \frac{1}{2}\sin\left(x-\frac{\pi}{3}\right)$$

Period: 2π

WORD OF ADVICE

If there is a minus sign in front of the constant c in a general sinusoidal equation, $y = a\sin(bx - c) + d$, the shifting goes right. If the sign in front of the constant c is a plus sign, the horizontal shifting goes left.

Solution: The graph of $y = \frac{1}{2}\sin\left(x - \frac{\pi}{3}\right)$ is obtained from the basic sine curve by vertically shrinking it in the y-direction and horizontally shifting it to the right.

Vertical Shift

The final type of transformation of the basic sine curve is the vertical shift caused by the constant d:

$$y = a\sin(bx - c) + d$$

The shift is d units upward if $d > 0$ and d units downward if $d < 0$. Then the curve oscillates about the horizontal line $y = d$ instead of oscillating about the x-axis.

Sample Problem 6

Sketch the graph of $y = 3\sin 2x + 2$.

Step 1: The amplitude is 3 and the y-coordinate of each key point is 3 times larger. This means that the graph is stretched vertically.

Step 2: Find the five key points. When b is a whole number, the period of the function is always less than 2π. In this case, b is 2, so the period is π—or two times less than 2π. This means that the x-coordinate of each key point should be divided by 2. Don't forget to change the y-coordinates as well (multiply by 3). Then, the key points become as follows:

$$(0,0), \left(\frac{\pi}{4},3\right), \left(\frac{\pi}{2},0\right), \left(\frac{3\pi}{4},-3\right), (\pi,0)$$

Step 3: Apply the vertical shift upward because d is positive. The graph is shifted 2 units up. Then the y-coordinates of key points increase by 2, and the key points become:

$$(0,2), \left(\frac{\pi}{4},5\right), \left(\frac{\pi}{2},2\right), \left(\frac{3\pi}{4},-1\right), (\pi,2)$$

Step 4: Sketch the graph using the new key points.

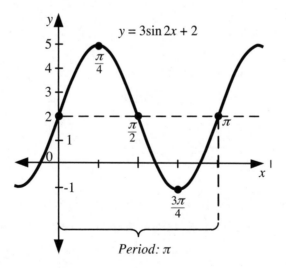

Solution: The graph of $y = 3\sin 2x + 2$ is obtained from the basic sine curve by horizontally shrinking it in the x-direction, vertically stretching it along the y-direction, and, finally, vertically shifting it 2 units upward.

The Cosine Curve

The cosine curve looks very much like the sine curve. It has the same domain (x-values, all real numbers) and the same range (y-values, from –1 to 1). The cosine curve also rolls without any interruption, repeating itself again and again.

For the cosine function, the key points are the following:

Maximum	Intercept	Minimum	Intercept	Maximum
$(0,1)$	$\left(\dfrac{\pi}{2},0\right)$	$(\pi,-1)$	$\left(\dfrac{3\pi}{2},0\right)$	$(2\pi,1)$
$(0,1)$	$(90°,0)$	$(180°,-1)$	$(270°,0)$	$(360°,1)$

Let's sketch the graph of the cosine function and compare it with the graph of the sine function. The cosine function is represented by the solid line; the sine function is the dashed one.

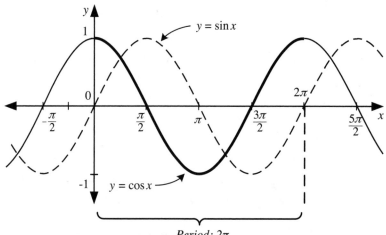

Period: 2π

If you look carefully at both graphs, you will notice that the cosine graph can be described as the sine graph shifted to the left by $\dfrac{\pi}{2}$ or 90°.

This means that everything previously discussed about stretching, shrinking, and shifting in relation to the sine graph holds true with the cosine graph as well, and you can apply the same techniques and rules when sketching graphs for the cosine function. Let's illustrate this point by solving the next problem.

But before you start solving the problem, let's look at the two graphs of the sine and cosine functions in the previous illustration. Notice that the sine graph is symmetrical with respect to the origin, whereas the cosine graph is symmetrical with respect to the y-axis. These symmetry properties attest to the fact that the sine function is odd whereas the cosine function is even. Both functions represent what all other odd and even functions do in terms of symmetry.

Sample Problem 7

Sketch the graph of $y = 3\cos(x + \pi)$.

Step 1: The amplitude is 3 and the y-coordinate of each key point is 3 times larger. This means that the graph is stretched vertically.

Step 2: Because c is positive, the horizontal shift of π goes left. Let's find the starting and ending points. Notice that because the shift goes left, the equations for shift phase have plus signs.

$$bx + c = 0 \rightarrow x + \pi = 0, \text{ thus } x = -\pi$$

$$bx + c = 2\pi \rightarrow x + \pi = 2\pi \rightarrow x = 2\pi - \pi, \text{ thus } x = \pi$$

So the first graph cycle starts at $-\pi$ and ends at π. To find x-coordinates for the other three key points, subtract π from each x-coordinate:

Second point: $\dfrac{\pi}{2} - \pi = \dfrac{\pi}{2} - \dfrac{2\pi}{2} = -\dfrac{\pi}{2}$

Third point: $\pi - \pi = 0$

Fourth point: $\dfrac{3\pi}{2} - \pi = \dfrac{3\pi}{2} - \dfrac{2\pi}{2} = \dfrac{\pi}{2}$

Then, the key points become:

$$(-\pi, 1), \ \left(-\dfrac{\pi}{2}, 0\right), \ (0, -1), \ \left(\dfrac{\pi}{2}, 0\right), \ (\pi, 1)$$

WORD OF ADVICE

When $c = \pm\pi$ in $y = a\sin(bx - c) + d$ and $y = a\cos(bx - c) + d$, the graphs of the sine and cosine functions are shifted left or right by π and look like upside-down versions of the original sine and cosine graphs. Flip the original graphs and you are done. Don't forget to change the period and the amplitude if $b \neq 1$ and $a \neq 1$.

Step 3: Change the y-coordinates (multiply by 3). Then, the key points become as follows:

$$(-\pi, 3), \left(-\frac{\pi}{2}, 0\right), (0, -3), \left(\frac{\pi}{2}, 0\right), (\pi, 3)$$

Step 4: Sketch the graph using the new key points.

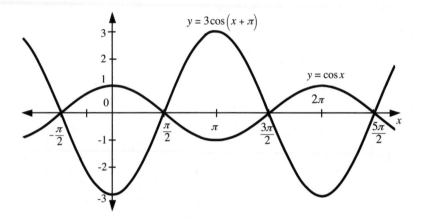

Solution: The graph of $y = 3\cos(x + \pi)$ is obtained from the basic cosine curve by vertically stretching it along the y-direction and horizontally shifting it by π to the left.

If you look closely at the graph that came as a result of the previous problem, you should notice that the graph looks like an upside-down version of the original cosine graph. This observation helps you sketch the graphs when the horizontal shift is π.

Modeling Periodic Behavior

As we discussed, the sinusoidal functions help to describe many natural phenomena and events from everyday life. For example, when a technician is tuning a piano and strikes a tuning fork for the A above middle C, he sets up a sound wave motion that can be approximated by the function $y = 0.001\sin880\pi t$, where t is time in seconds.

Sales S (in thousands of units) over the period of time t (in months) are often presented by either the sine or cosine functions such as:

$$S = 22.5 - 3.4\cos\frac{\pi t}{6}$$

Blood pressure and respiratory cycle can be modeled by these trigonometric functions as well. The last problem you do in this chapter is related to your health.

Sample Problem 8

The following function $P = 100 - 20\cos\dfrac{5\pi t}{3}$ describes the blood pressure P (where t is time in seconds) for a person at rest. The periodic fluctuation of the pressure is due to the heart pumping the blood around the body. Find the period of the function and the number of heartbeats per minute.

Step 1: To find the period of the function, divide 2π by the frequency constant $b = \dfrac{5\pi}{3}$:

$$2\pi \div \frac{5\pi}{3} = \frac{2\pi}{1} \div \frac{5\pi}{3} = \frac{2\pi}{1} \cdot \frac{3}{5\pi} = \frac{6}{5}$$

The period is $\dfrac{6}{5}$ seconds.

Step 2: Because the period is $\dfrac{6}{5}$ seconds, in order to find the number of heartbeats per minute, let's first find the number per each second since that's the unit that the period is expressed in. The number of heartbeats per second is $\dfrac{6}{5}$, which is the reciprocal fraction of the period. Since you now know the number of heartbeats per second, the number of heartbeats per minute (60 seconds) can be calculated as:

$$\frac{5}{6} \cdot 60 = 50$$

Solution: The period of the function is $\dfrac{6}{5}$ and the number of heartbeats per minute is 50.

As the preceding problem illustrates, the sine and cosine functions are very useful to describe and model real-life situations.

Practice Problems

Problem 1: Sketch the graph of $y = 2\sin x$.

Problem 2: Sketch the graph of $y = \dfrac{1}{2}\cos x$.

Problem 3: Sketch the graph of $y = 2\sin\dfrac{x}{4}$.

Problem 4: Sketch the graph of $y = 3\cos(x + \pi)$.

Problem 5: Sketch the graph of $y = -\dfrac{1}{2}\sin\dfrac{x}{2} + 3$.

The Least You Need to Know

- The general equations for the sine and cosine functions are $y = a\sin(bx - c) + d$ and $y = a\cos(bx - c) + d$.
- The amplitude a tells you how far from 0 the values of the function can get. The curves are stretched or shrunk along the y-axis.
- The frequency b tells you how many periods are repeated in an interval of 2π. The period of the function is $\dfrac{2\pi}{b}$. The curves are stretched or shrunk along the x-axis.
- The phase shift c tells you how much the curve has been shifted left or right.
- The curves of functions also can be shifted along the y-axis by the term d at the end of the right side of the general equation.

Graphs of Other Trigonometric Functions

In This Chapter

- Finding asymptotes
- Exploring effects of multiplication and addition
- Comparing the tangent and cotangent
- Using the sine and cosine graphs to graph their reciprocals

In the previous chapter we dealt with shifting, stretching, and shrinking of the sinusoidal graphs. This chapter introduces the graphs of the other four trigonometric functions.

Graphs of Tangent Function

The graph of the tangent function is different from the sine and cosine graphs in two significant ways.

First, the domain is different. Recall that the domain (*x*-values) for both the sine and cosine functions is all real numbers; therefore, the graphs roll uninterrupted from negative infinity to positive infinity. Because we identify the tangent as a ratio $\tan \theta = \dfrac{\sin \theta}{\cos \theta}$, it is clear that the tangent is undefined when the cosine is equal to 0, so the graph will be interrupted.

Second, the basic sine and cosine functions are bounded in a sense that the values they take are always between –1 and 1 (inclusive). But the tangent function takes on all real numbers as *y*-values.

But let's discuss everything in turn.

Domain and Period of the Tangent

Because the tangent is undefined when $\cos\theta = 0$, you need to exclude all the values for angles whose cosines are equal to 0. Two such values are $\pm 90°$ or $\pm\dfrac{\pi}{2}$ because $\cos\left(\pm\dfrac{\pi}{2}\right) = 0$. Therefore, all angles such as $\dfrac{\pi}{2} + n\pi$, where n is an integer, must be excluded from the domain of the tangent function.

Thus, the domain of the tangent is:

$$x \neq \frac{\pi}{2} + n\pi$$

It is clear from the previous discussion that the period of the tangent is π, because after every π (180°), you encounter an angle whose cosine is equal 0. Therefore, the period of the tangent is π.

As x goes from $-\dfrac{\pi}{2}$ to $\dfrac{\pi}{2}$, the tangent takes on all real values. Thus:

$$-\frac{\pi}{2} < x < \frac{\pi}{2}$$

$$-\infty < \tan x < +\infty$$

Quadrants I and IV constitute a complete period of $y = \tan x$. In quadrant I the tangent is positive, and in quadrant IV it is negative.

Asymptotes

At the quadrant angles $-\dfrac{\pi}{2}$ and $+\dfrac{\pi}{2}$ the tangent is undefined, as we discussed before, so the graph gets interrupted and then picks up after the quadrant angles.

When x (angle value) approaches $\dfrac{\pi}{2}$, $\tan x$ increases without bound (because you divide by 0), so the range (y-values) of the tangent is all real numbers (from minus infinity to plus infinity).

Let's sketch the graph of the tangent and discuss it further.

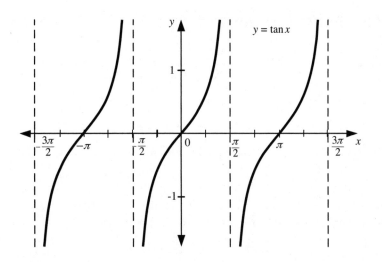

Note that the graph approaches lines $x = \dfrac{\pi}{2} + n\pi$ but never reaches them. These lines are called *vertical asymptotes* of the curve $y = \tan x$. By replacing n with various integers in the equation for the asymptotes, you get lines such as:

$$-\frac{7\pi}{2}, \ -\frac{5\pi}{2}, \ -\frac{3\pi}{2}, \ -\frac{\pi}{2}, \ \frac{\pi}{2}, \ \frac{3\pi}{2}, \ \frac{5\pi}{2}, \ \frac{7\pi}{2}$$

DEFINITION

A **vertical asymptote** is a line that a curve approaches more and more closely but never reaches.

Obviously, this list can be continued infinitely in both directions.

Altering the Tangent Graph

You can change the graph of the tangent function similar to the ways you change the sine and cosine functions with addition and multiplication, but the effects of these alterations will be different.

Let's again consider a general form of the tangent function and discuss everything in turn:

$$y = a\tan(bx - c) + d$$

Altering the Tangent with Multiplication

The amplitude of a tangent is not defined. Multiplying the tangent function by a number a is not going to change it the way it changes the sine and cosine functions because the tangent values already go from minus infinity to plus infinity. However, multiplication of the tangent function by a number affects the speed with which the graph becomes steeper or flatter. This is how it works:

- Multiplying by a number greater than 1:

  When multiplied by a number greater than 1, the function gets steeper more quickly. For example, if the angle is $30°\left(\dfrac{\pi}{6}\right)$, then $\tan\dfrac{\pi}{6} = \dfrac{\sqrt{3}}{3} \approx 0.58$, but $2\tan\dfrac{\pi}{6} \approx 2 \cdot 0.58 \approx 1.15$, so the values for y become larger twice as fast as for the basic tangent function.

- Multiplying by a fraction:

  When multiplied by a fraction, the function gets flatter. For example, if the angle is again $30°\left(\dfrac{\pi}{6}\right)$, then $\tan\dfrac{\pi}{6} = \dfrac{\sqrt{3}}{3} \approx 0.58$, but $\dfrac{1}{2}\tan\dfrac{\pi}{6} \approx 0.29$, so the y-values increase more slowly.

- Multiplying by a negative number:

  When multiplied by a negative number, the tangent curve flips upside down over the x-axis.

• The following illustration depicts all the discussed changes.

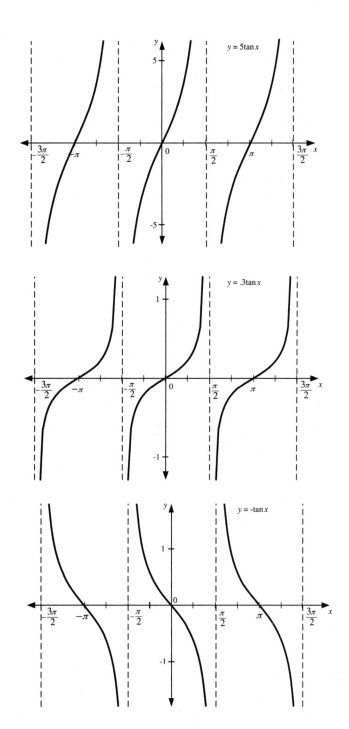

Altering the Tangent with Multiplication of the Angle

Multiplication of the angle affects the tangent graph the same way this operation affected the sine and cosine graphs—it changes the period. For example, $\tan 2x$ makes two cycles over the usual period of π, so the period of $\tan 2x$ is $\frac{\pi}{2}$. On the contrary, the graph $\tan \frac{x}{2}$ has a period of 2π; thus it has only half as many cycles over the usual period, and it takes twice as long to complete a full cycle.

The period of the tangent function is the distance between two asymptotes. In general, to find the asymptotes, solve the equations:

$$bx - c = -\frac{\pi}{2} \text{ and } bx - c = \frac{\pi}{2}$$

Let's illustrate how it works with an example.

Sample Problem 1

Sketch the graph of $y = \tan \frac{x}{4}$.

Step 1: Find the asymptotes by solving the equations, using the values from the function $c = 0$ and $b = \frac{1}{4}$:

$$bx - c = -\frac{\pi}{2} \;\rightarrow\; \frac{1}{4}x = -\frac{\pi}{2} \;\rightarrow\; x = -\frac{\pi}{2} \div \frac{1}{4} = -\frac{\pi}{2} \cdot \frac{4}{1} = -2\pi$$

$$bx - c = \frac{\pi}{2} \;\rightarrow\; \frac{1}{4}x = \frac{\pi}{2} \;\rightarrow\; x = \frac{\pi}{2} \div \frac{1}{4} = \frac{\pi}{2} \cdot \frac{4}{1} = 2\pi$$

The period of the function is 4π (from -2π to $+2\pi$) with asymptotes at -2π and 2π for the main cycle.

Step 2: Sketch the graph using the information from Step 1.

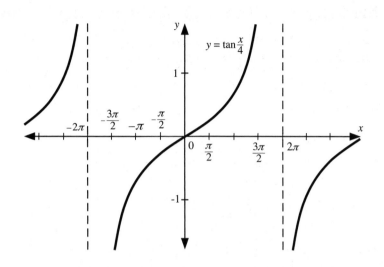

Solution: The graph is a basic tangent graph stretched along x-axis.

Altering the Tangent with Addition

Similar to the sine and cosine graphs, adding numbers shifts the graph.

When you add a number to or subtract a number from the angle, the tangent graph shifts either to the left or to the right along the x-axis. For example, the graph $\tan\left(x - \dfrac{\pi}{4}\right)$ shifts $\dfrac{\pi}{4}$ radians to the right.

DANGEROUS TURN

When shifting, be careful with the direction of shifting. Note that when you add to the angle, the tangent graph shifts to the left, and when you subtract from the angle, the tangent graph shifts to the right.

Finally, when a number is added to or subtracted from the tangent function, it results in a vertical shift of the graph. Adding the number shifts the graph upward along the y-axis, whereas subtracting the number shifts the graph downward along the y-axis.

Recall that in the sine and cosine graphs the maximum and the minimum points of the graphs changed the values with this sort of addition. Because the values of the tangent function are from negative infinity to positive infinity, adding or subtracting numbers does not change the values of the function. What changes is the *point of inflection* in the tangent curve.

DEFINITION

The **point of inflection** is a point on a curve at which the curvature changes sign. The curve changes from being concave upward (positive curvature) to concave downward (negative curvature) or vice versa.

The following illustration demonstrates the discussed shifts.

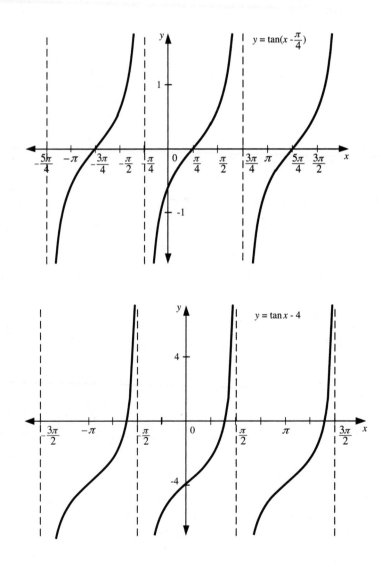

Graphs of Cotangent Function

The graph of the cotangent function is similar to the graph of the tangent function in many ways. Its range is from negative infinity to positive infinity and its period is π as well.

The general form of the cotangent function is similar to the one for the tangent:

$$y = a\cot(bx - c) + d$$

Graphs of both functions have asymptotes. However, by definition $\cot x = \dfrac{\cos x}{\sin x}$; therefore, the cotangent is undefined for all angles where the sine is equal to 0. Thus, the cotangent function has vertical asymptotes at $x = n\pi$. Some of these asymptotes are:

$$-3\pi,\ -2\pi,\ -\pi,\ 0,\ \pi,\ 2\pi,\ 3\pi$$

Also, because these two functions, the tangent and cotangent, are reciprocals, the direction of their rise and fall is different. The values of the tangent function increase as you go from left to right, but the values of the cotangent function decrease as you go from left to right.

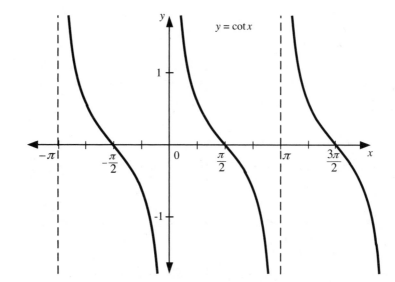

Sample Problem 2

Sketch the graph of $y = -2\cot\dfrac{x}{3} + 2$.

Step 1: Let's first analyze what the first constant in the equation means. While the 2 in front of the cotangent is not formally defined as amplitude since the cotangent doesn't have maximum or minimum points, the constant 2 does affect the steepness of the graph. Also, because you have a minus sign in front of the 2, the graph flips over the x-axis.

Step 2: The asymptotes for the basic cotangent curve are 0 and π. Let's apply equations to solve for these asymptotes:

$$bx - c = 0 \text{ and } bx - c = \pi$$

Using the given values from the function $c = 0$ and $b = \dfrac{1}{3}$:

$$bx - c = 0 \rightarrow \frac{1}{3}x = 0 \rightarrow x = 0 \div \frac{1}{3} = 0$$

$$bx - c = \pi \rightarrow \frac{1}{3}x = \pi \rightarrow x = \frac{\pi}{1} \div \frac{1}{3} = \frac{\pi}{1} \cdot \frac{3}{1} = 3\pi$$

Because the period for the cotangent is the distance between two asymptotes, you can state that the period in this case is 3π.

WORD OF ADVICE

To find the starting and ending points of the main cycle of trigonometric functions when the angle is multiplied by a number, solve the following equations:

$$bx - c = p_1$$

$$bx - c = p_2$$

where p_1 and p_2 are the starting and ending points of the main cycle of the basic graph.

These new points are asymptotes for the functions that have them. The distance between these new points of the cycle are the period of the altered trigonometric function.

Step 3: Determine the vertical shift. The positive number is added to the function, so the graph shifts up 2 units. Sketch the graph of the function.

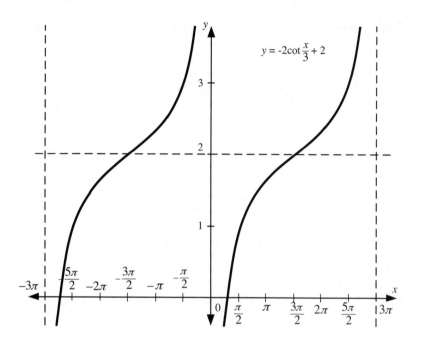

$$y = -2\cot\frac{x}{3} + 2$$

Solution: The graph of $y = -2\cot\dfrac{x}{3} + 2$ is a flipped graph of a basic cotangent graph that is stretched along the x-axis, shifted up 2 units along the y-axis, and steeper due to multiplication of the function by 2.

Graphs of the Reciprocal Functions

There are two functions left to discuss—the cosecant and the secant. They are reciprocals of the sine and cosine functions; therefore, the easiest way to sketch their graphs is to relate them to the sine and cosine graphs.

Graph of the Cosecant Function

The cosecant function is the reciprocal of the sine function, and this fact affects the domain of the cosecant function. Whenever the sine function is equal to 0, the cosecant function doesn't exist; therefore, you can determine the asymptotes of the cosecant. Because the sine function is equal to 0 for all angles that are multiples of π, the asymptotes of the cosecant function are of the form $x = n\pi$, where n is an integer. Some of the examples for asymptotes are:

$$-3\pi,\ -2\pi,\ -\pi,\ 0,\ \pi,\ 2\pi,\ 3\pi$$

The easiest way to sketch the cosecant graph is to sketch the sine function as an aid and draw the asymptotes through the x-values where the sine graph intercepts the x-axis.

The second important aid comes from the maximum and minimum y-values of the sine graph. They identify the points where the cosecant graph reverses its slope.

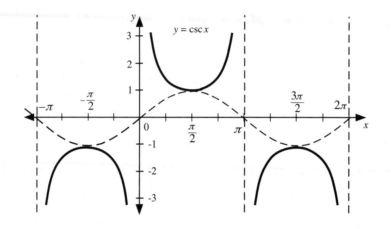

We summarize the discussion by stating that the period of the cosecant function is 2π, the domain is all $x \neq n\pi$, and the range is all y that are outside of $(-1,1)$. The asymptote equation is $x = n\pi$. Note that with the two graphs, the hills and valleys are interchanged. When the sine curve reaches a hill (a local maximum), the cosecant graph reaches the valley (the local minimum), and vice versa. The cosecant function is an odd function because it is a reciprocal of the sine function, and it is symmetrical with respect to the origin. As with all trigonometric functions, the cosecant curve repeats itself over and over again.

Changes to the cosecant function can be made in a way similar to the other trigonometric functions that we already discussed.

Adding to or subtracting from the angle results in horizontal shift of the basic cosecant graph along the x-axis. You can predict that, as with other trig functions, adding to or subtracting from the cosecant function results in a vertical shift of the graph along y-axis. The following graph shows what $y = \csc\left(x + \dfrac{\pi}{2}\right) + 3$ looks like.

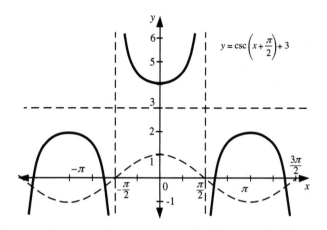

Multiplying the function of the cosecant changes the steepness of the graph. Multiplying the angle changes the period of the graph. The rule is the same as with other trig functions. When the angle is multiplied by a whole number, the graph shrinks horizontally because more periods fit into the original period. When the angle is multiplied by a fraction, the graph stretches horizontally along the x-axis. The following illustration depicts the changes for $y = 2\csc\dfrac{x}{2}$ in comparison to the basic sine curve depicted as a dotted line.

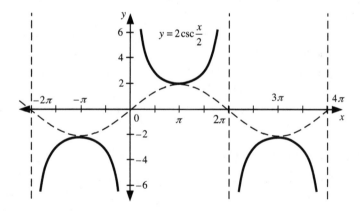

Observe that the graph becomes steeper than the basic cosecant graph (the units of the y-axis scale have increased).

Graph of the Secant Function

The secant function is the reciprocal of the cosine function, and this fact affects the domain of the secant function; therefore, whenever the cosine function is equal to 0, the secant function doesn't exist. You can determine the asymptotes of the secant by recalling that the cosine function is equal to 0 for angles $\frac{\pi}{2} + n\pi$. Thus, the asymptotes of the secant function are of the form $x = \frac{\pi}{2} + n\pi$, where n is an integer.

The easiest way to sketch the secant graph is to sketch the cosine function and draw the asymptotes through the x-values where the cosine graph intercepts the x-axis.

WORTH KNOWING

Note that both the secant and cosecant functions have no amplitude and their basic graphs have no x-intercepts. Meanwhile, the tangent and the cotangent functions also do not have amplitudes, but always have x-intercepts.

The second important aid comes, as with the cosecant function, from the maximum and minimum y-values of the cosine graph. They identify the points where the secant graph changes slope.

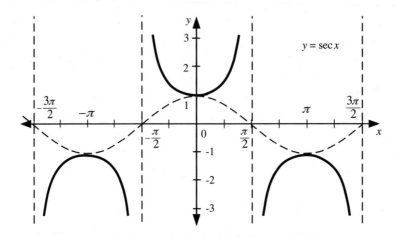

We summarize the discussion by stating that the period of the secant function is 2π, the domain is all $x \neq \frac{\pi}{2} + n\pi$, and the range is all y not in $(-1,1)$. The asymptote equation is $x = \frac{\pi}{2} + n\pi$. The hills and valleys are interchanged similar to the sine and cosecant functions. When the cosine curve reaches a hill (a local maximum), the secant graph reaches the valley (the local minimum), and vice versa. The secant function is an even function because it is a reciprocal of the cosine function, and it is symmetrical with respect to the y-axis. As with all trigonometric functions, the secant curve repeats itself over and over again.

Changes to the secant function can be made in a way similar to the cosecant function and all the other trigonometric functions we've already discussed. As an example, the following is a graph of $y = \frac{1}{5}\sec 2x$.

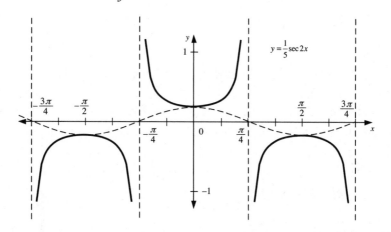

Notice that in comparison to the basic secant graph, the graph in the illustration is flattened and much closer to the x-axis. This is the result of multiplying the function by a fraction. Also, the period is two times shorter, and some of the asymptotes are:

$$-\frac{\pi}{4}, \frac{\pi}{4}, \frac{3\pi}{4}, \frac{5\pi}{4}$$

Graphing Combinations of Functions

Sums, differences, products, and quotients of trigonometric functions are also periodic. The period of the function resultant from a combination of several component functions is equal to the least common multiple of the periods of the component functions. This period is important for determining an appropriate viewing rectangle for the function on the graphing calculator, which we will go over in Chapter 22 of this book. Some examples are:

$y = \sin x - \cos 2x$

$y = 3\sin x + 2\sin 4x$

Combinations of algebraic functions and trigonometric functions are generally not periodic. Some examples are:

$$y = -x + \sin x$$

$$y = x^2 - \cos x$$

You might wonder why you even need to consider these functions and whether they have any relations to real-life situations, or whether they have been created for the sheer amusement of professional mathematicians. It turns out that many applications in physics and biology can be described by curves that are results of addition or subtraction of two trigonometric functions or one algebraic and one trigonometric function.

We discuss how to sketch the graphs of some of these functions using the graphing calculator in Chapter 22 of this book.

Practice Problems

Problem 1: Sketch the graph of $2\tan\dfrac{x}{3}$.

Problem 2: Sketch the graph of $\dfrac{1}{3}\cot x + 2$.

Problem 3: Sketch the graph of $-3\tan 2x$.

Problem 4: Sketch the graph of $3\csc\left(x + \pi\right)$.

Problem 5: Sketch the graph of $\dfrac{1}{2}\sec 2x$.

The Least You Need to Know

- The range for the tangent and cotangent is all real numbers; the range for secant and cosecant is all values of y not in $(-1,1)$.
- The period of the tangent and cotangent functions is π.
- The period of the secant and cosecant functions is 2π.
- Vertical asymptotes for tangent and secant are $x = \dfrac{\pi}{2} + n\pi$.
- Vertical asymptotes for cotangent and cosecant are $x = n\pi$.

Graphs of Inverse Trigonometric Functions

In This Chapter

- Finding the domain and range of inverse trig functions
- Evaluating inverse trig functions
- Finding angles that the inverse properties are applied for
- Using right triangles to find values of inverse functions

Corresponding to each trigonometric function, there is an inverse trigonometric function. If you need to review inverse functions, refer to Chapter 3. In this chapter, we explore the graphs of the inverse trigonometric functions.

Recall that there are two possible notations for inverse trig functions. The notation $\sin^{-1}x$ is consistent with the general inverse function notation. The arcsin$x$ notation (arcsine of x) comes from the association of a central angle and the corresponding arc on the unit circle. Both notations are common in mathematics.

If you try to translate arcsin$x$ into plain English, it would be: "The angle whose sine is x." For example, the equation $\arcsin\frac{1}{2} = \frac{\pi}{6}$ can be translated as: "The angle whose sine is $\frac{1}{2}$ is $\frac{\pi}{6}$." Generally speaking, with inverse trigonometric functions, you are looking for the angle whose given trigonometric function has a specific value. In an expression arctan1, you are looking for the angle whose tangent value is equal to 1. As discussed in previous chapters, this angle happens to be $\frac{\pi}{4}$.

Inverse Sine Function

If you look at the graph of the sine function, it is clear it would not pass the horizontal line test for inverse functions, because different values of x yield the same y-values.

However, if you restrict the domain and consider the sine function in the interval $-\frac{\pi}{2} \leq x \leq \frac{\pi}{2}$, the sine function has one-to-one correspondence of the x- and y-values on the limited interval.

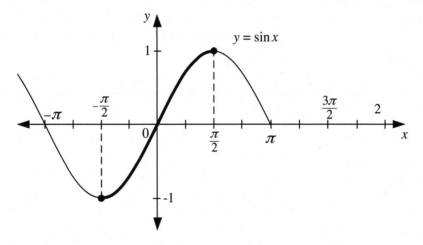

Thus, on the domain that is restricted to $-\frac{\pi}{2} \leq x \leq \frac{\pi}{2}$, which is depicted by the bold line in the previous figure, the sine function has a unique *inverse sine function*, which is denoted by $y = \sin^{-1}x$ or $y = \arcsin x$.

DEFINITION

The **inverse sine function** is defined by $y = \arcsin x$ if and only if $\sin y = x$, where $-1 \leq x \leq 1$ and $-\frac{\pi}{2} \leq y \leq \frac{\pi}{2}$. The domain of $y = \arcsin x$ is $[-1,1]$, and the range is $\left[-\frac{\pi}{2}, \frac{\pi}{2} \right]$.

When dealing with the inverse sine function, it is helpful to remember this sentence: "The arcsine of x is the angle whose sine is x."

Sketching the Graph of Arcsine Function

On the interval $-\dfrac{\pi}{2} \le y \le \dfrac{\pi}{2}$, the equations $y = \arcsin x$ and $\sin y = x$ are equivalent; hence, their graphs are the same. Let's try to understand the equivalency of these two equations one more time with the plain English:

$y = \arcsin x$ means "The radian angle y is the angle whose sine is x."

$\sin y = x$ means "The radian angle y has the sine whose value is x."

The fact of equivalency of these two equations helps you to sketch the graph for arcsine x by assigning values to y in the second equation. Let's construct the following table of values and sketch a graph of the inverse sine using the values from the table.

y	$-\dfrac{\pi}{2}$	$-\dfrac{\pi}{4}$	$-\dfrac{\pi}{6}$	0	$\dfrac{\pi}{6}$	$\dfrac{\pi}{4}$	$\dfrac{\pi}{2}$
$x = \sin y$	-1	$-\dfrac{\sqrt{2}}{2}$	$-\dfrac{1}{2}$	0	$\dfrac{1}{2}$	$\dfrac{\sqrt{2}}{2}$	1

The resulting graph is shown in the following figure.

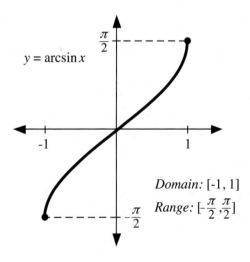

$y = \arcsin x$

Domain: [-1, 1]

Range: $[-\dfrac{\pi}{2}, \dfrac{\pi}{2}]$

Be sure that you understand that the illustration shows the entire graph of the inverse sine function and not just part of it, because its domain and range are restricted by the values between −1 and +1 and between $-\dfrac{\pi}{2}$ and $\dfrac{\pi}{2}$, as we discussed previously.

Evaluating the Inverse Sine Function

To evaluate the inverse sine function, you use the definition of the inverse sine function.

Sample Problem 1

Find the values of the following: $\arcsin\dfrac{\sqrt{2}}{2}$ and $\sin^{-1}\left(-\dfrac{\sqrt{3}}{2}\right)$.

Step 1: By definition, $y = \arcsin\dfrac{\sqrt{2}}{2}$ implies that $\sin y = \dfrac{\sqrt{2}}{2}$. On the interval $-\dfrac{\pi}{2} \le y \le \dfrac{\pi}{2}$, there is only one angle, $\dfrac{\pi}{4}$, with a sine equal to $\dfrac{\sqrt{2}}{2}$. Thus, you can conclude the following:

$$\arcsin\dfrac{\sqrt{2}}{2} = \dfrac{\pi}{4}$$

DANGEROUS TURN

It is impossible to evaluate inverse function $\sin^{-1}3$ because there is no angle with a sine of 3. Recall that the domain of the inverse sine function is [−1,1].

Step 2: By definition, $y = \sin^{-1}\left(-\dfrac{\sqrt{3}}{2}\right)$ implies that $\sin y = -\dfrac{\sqrt{3}}{2}$. On the interval $-\dfrac{\pi}{2} \le y \le \dfrac{\pi}{2}$, there is only one angle, $-\dfrac{\pi}{3}$, with a sine equal to $-\dfrac{\sqrt{3}}{2}$. Thus, you can conclude the following:

$$\sin^{-1}\left(-\dfrac{\sqrt{3}}{2}\right) = -\dfrac{\pi}{3}$$

Solution: You find that $\arcsin\dfrac{\sqrt{2}}{2} = \dfrac{\pi}{4}$ and $\sin^{-1}\left(-\dfrac{\sqrt{3}}{2}\right) = -\dfrac{\pi}{3}$.

Other Inverse Trigonometric Functions

This section summarizes other inverse trigonometric functions. Here, we discuss in more detail the other two most common inverse trigonometric functions—the inverse cosine and the inverse tangent. The inverses for the reciprocal functions are briefly mentioned as well.

Inverse Cosine and Tangent Functions

The cosine function is uniquely defined (in the sense that it does not repeat itself) on the interval $0 \le x \le \pi$. Consequently, on this interval the cosine has an inverse function denoted as $y = \arccos x$ or $y = \cos^{-1} x$. Because the cosine can be equal to values only from –1 to +1, the domain of the inverse cosine (i.e., $\cos^{-1}$) is $-1 \le x \le 1$ and the range is $0 \le y \le \pi$.

The following illustration shows the cosine function on the interval where it is unique as well as the inverse cosine function.

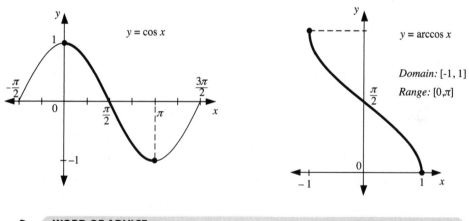

WORD OF ADVICE

Note that when you sketch the graphs of the inverse trigonometric functions, in contrast with the graphs of trigonometric functions, the y-values represent the angles in radians and x-values are numbers.

Notice that the x-values are the same for both the inverse sine and inverse cosine, but the y-values are different. The y-values (or the range) for the inverse sine are angles from the first and fourth quadrants $-\dfrac{\pi}{2} \le y \le \dfrac{\pi}{2}$, whereas the y-values (or the range) for the inverse cosine are angles from the first and second quadrants $0 \le y \le \pi$.

Similarly, we can define an inverse tangent function of $\tan^{-1}x$. Let's start off with the tangent function, which is uniquely defined on the interval $-\dfrac{\pi}{2} < x < \dfrac{\pi}{2}$. Consequently, on this interval the tangent has an inverse function denoted as $y = \arctan x$ or $y = \tan^{-1}x$. Because the tangent can be equal to values from negative infinity to positive infinity, the domain of the inverse tangent is $-\infty < x < +\infty$ and the range is $-\dfrac{\pi}{2} < y < \dfrac{\pi}{2}$.

The following illustration shows the tangent function on the interval where it is unique as well as the inverse tangent function.

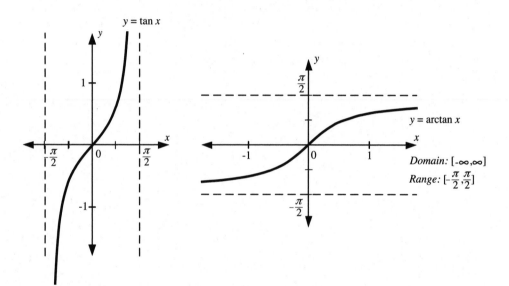

Note that the graph of the inverse tangent has two horizontal asymptotes, $y = -\dfrac{\pi}{2}$ and $y = \dfrac{\pi}{2}$, because the tangent function does not exist for these two angles. You can recall that the tangent function is not defined whenever the cosine is equal to zero, which is the case with these two angles.

In summary, the graph of the inverse tangent has the x-values from negative infinity to positive infinity, and the y-values that are between the two asymptotes $-\dfrac{\pi}{2}$ and $\dfrac{\pi}{2}$.

Inverses of Reciprocal Trigonometric Functions

In this section, we introduce the inverse functions of the three reciprocal functions. They are not as commonly used as the three that were discussed in previous sections, but it is still useful to know them.

The cotangent function is uniquely defined on the interval $0 < x < \pi$. Consequently, on this interval the cotangent has an inverse function denoted as $y = \text{arccot}\, x$ or $y = \cot^{-1} x$. Because the cotangent can be equal to values from negative infinity to positive infinity, the domain of the inverse cotangent is $-\infty < x < +\infty$ and the range is $0 < y < \pi$.

The following illustration shows the inverse cotangent function.

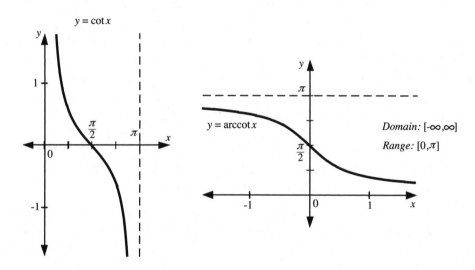

Note that the graph of the inverse cotangent has two horizontal asymptotes, $y = 0$ and $y = \pi$, because the cotangent function does not exist for these two angles. You can recall that the cotangent function is not defined whenever the sine is equal to zero, which is the case with the two previously mentioned angles.

The main difference between the inverse tangent and inverse cotangent curves is that the first rises as you go from left to right and the second falls as you go from left to right.

The graphs of inverse cosecant and inverse secant are probably the most elaborate among the inverse function curves.

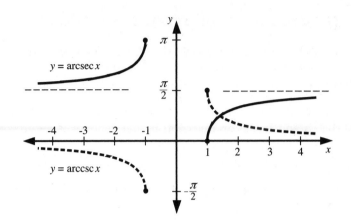

The first peculiarity about these graphs is that they have a void in the middle of their graphs. This is because their parent functions, the secant and cosecant, don't have y-values between -1 and 1.

The domain (x-values) of $y = \text{arccsc}\, x$ is from negative infinity to -1 (up to and including it) and from 1 (also inclusive of it) to positive infinity. The range (y-values) of the inverse cosecant is $-\dfrac{\pi}{2} \le y < 0$ and $0 < y \le \dfrac{\pi}{2}$.

The x-axis is the horizontal asymptote of the inverse cosecant graph because the cosecant function is not defined for the angle of zero.

The domain (x-values) of $y = \text{arcsec}\, x$ is from negative infinity to -1 (inclusive) and from 1 (inclusive) to positive infinity. The range (y-values) of the inverse secant is $0 \le y < \dfrac{\pi}{2}$ and $\dfrac{\pi}{2} < y \le \pi$.

The line $y = \dfrac{\pi}{2}$ is the horizontal asymptote of the inverse secant graph because the secant function is not defined for the angle of $\dfrac{\pi}{2}$.

Using the information you acquired in the previous sections, let's do several sample problems.

Sample Problem 2

Find the exact values of the following: $\arccos\dfrac{\sqrt{2}}{2}$, $\cos^{-1}(-1)$, and $\arccos\left(-\dfrac{\sqrt{3}}{2}\right)$.

Step 1: Because $\cos\dfrac{\pi}{4} = \dfrac{\sqrt{2}}{2}$ and $\dfrac{\pi}{4}$ lies in the range of the inverse cosine $[0, \pi]$, you can conclude the following:

$$\arccos\dfrac{\sqrt{2}}{2} = \dfrac{\pi}{4}$$

Step 2: Because $\cos\pi = -1$ and π lies in the range of the inverse cosine $[0, \pi]$, you can conclude the following:

$$\cos^{-1}(-1) = \pi$$

Step 3: Because $\cos\left(\dfrac{5\pi}{6}\right) = -\dfrac{\sqrt{3}}{2}$ and $\dfrac{5\pi}{6}$ lies in the range of the inverse cosine $[0, \pi]$, you can conclude the following:

$$\arccos\left(-\dfrac{\sqrt{3}}{2}\right) = \dfrac{5\pi}{6}$$

Solution: You find that $\arccos\dfrac{\sqrt{2}}{2} = \dfrac{\pi}{4}$, $\cos^{-1}(-1) = \pi$, and $\arccos\left(-\dfrac{\sqrt{3}}{2}\right) = \dfrac{5\pi}{6}$.

WORD OF ADVICE

Some of the inverse functions are nearly impossible to approximate without the use of a graphing calculator. Refer to the Chapter 22 to learn how use the functions of the graphing calculator.

The next problem involves inverse tangents.

Sample Problem 3

Find the exact values of the following: $\arctan 0$, $\tan^{-1}\left(-\sqrt{3}\right)$, and $\arctan\left(-\dfrac{\sqrt{3}}{3}\right)$.

Step 1: Because $\tan 0 = 0$ and 0 lies in the range of the inverse tangent $\left[-\dfrac{\pi}{2}, \dfrac{\pi}{2}\right]$, you can conclude the following:

$$\arctan 0 = 0$$

Step 2: Because $\tan\left(-\dfrac{\pi}{3}\right) = -\sqrt{3}$ and $-\dfrac{\pi}{3}$ lies in the range of the inverse tangent $\left[-\dfrac{\pi}{2}, \dfrac{\pi}{2}\right]$, you can conclude the following:

$$\tan^{-1}\left(-\sqrt{3}\right) = -\dfrac{\pi}{3}$$

Step 3: Because $\tan\left(-\dfrac{\pi}{6}\right) = -\dfrac{\sqrt{3}}{3}$ and $-\dfrac{\pi}{6}$ lies in the range of the inverse tangent $\left[-\dfrac{\pi}{2}, \dfrac{\pi}{2}\right]$, you can conclude the following:

Solution: You find that $\arctan 0 = 0$, $\tan^{-1}\left(-\sqrt{3}\right) = -\dfrac{\pi}{3}$, and $\arctan\left(-\dfrac{\sqrt{3}}{3}\right) = -\dfrac{\pi}{6}$.

Compositions of Trig and Inverse Trig Functions

Recall from Chapter 3 that all inverse functions possess the following properties:

$$f\left(f^{-1}(x)\right) = x \quad \text{and} \quad f\left(f^{-1}(y)\right) = y$$

WORTH KNOWING

If $-1 \le x \le 1$ and $-\dfrac{\pi}{2} \le y \le \dfrac{\pi}{2}$, then $\sin(\arcsin x) = x$ and $\arcsin(\sin y) = y$.

If $-1 \le x \le 1$ and $-0 \le y \le \pi$, then $\cos(\arccos x) = x$ and $\arccos(\cos y) = y$.

If $-\dfrac{\pi}{2} < y < \dfrac{\pi}{2}$, then $\tan(\arctan x) = x$ and $\arctan(\tan y) = y$.

The two properties stated above imply that a regular function nested within its inverse function cancels itself out. Let's go over this situation in more detail and understand it based on an example.

We again translate the language of math into plain English.

The expression $\sin\dfrac{\pi}{6} = \dfrac{1}{2}$ means, "The sine of the angle $\left(\dfrac{\pi}{6}\right)$ is $\dfrac{1}{2}$." Let's look at an analogy from a situation that is not related to math. If you use the word *name* as an analogy for the term *sine*, the expression *this particular girl* for the angle of $\left(\dfrac{\pi}{6}\right)$, and *Anna* as an analogy for the sine value of $\dfrac{1}{2}$, then you can say, "The name of this particular girl is Anna."

The expression $\arcsin\dfrac{1}{2}=\dfrac{\pi}{6}$ means, "The angle whose sine is $\dfrac{1}{2}$ is angle $\left(\dfrac{\pi}{6}\right)$." With this analogy, you can say, "The girl, whose name is Anna, is this particular girl."

The expression $\sin\left(\arcsin\dfrac{1}{2}\right)=\dfrac{1}{2}$ means, "The sine of the angle whose sine is $\dfrac{1}{2}$ is equal to $\dfrac{1}{2}$." If you again use the previous analogy, you can say, "The name of this particular girl, whose name is Anna, is still Anna." Obviously, the answer is already in the statement.

We now find the exact values of some functions using inverse properties.

Sample Problem 4

Find the exact values of $\tan(\arctan(-7))$, $\arcsin\left(\sin\dfrac{5\pi}{3}\right)$, and $\cos\left(\cos^{-1}2\pi\right)$.

Step 1: Because -7 is in the domain of arctan x, you can apply the inverse property:

$$\tan(\arctan(-7)) = -7$$

Step 2: Because the range of the inverse sine function is $-\dfrac{\pi}{2}\le y\le\dfrac{\pi}{2}$, the angle $\dfrac{5\pi}{3}$ does not lie within this range. Nevertheless, $\dfrac{5\pi}{3}$ is coterminal with another angle that belongs to the range and has the same sine value as the original angle $\dfrac{5\pi}{3}$:

$$\dfrac{5\pi}{3}-2\pi=\dfrac{5\pi}{3}-\dfrac{6\pi}{3}=-\dfrac{\pi}{3}$$

Thus, you can rewrite the problem as follows:

$$\arcsin\left(\sin\dfrac{5\pi}{3}\right)=\arcsin\left(\sin\left(-\dfrac{\pi}{3}\right)\right)=-\dfrac{\pi}{3}$$

DANGEROUS TURN

Inverse properties cannot be applied to any arbitrary values of x and y. For example, $\arcsin\left(\sin\dfrac{3\pi}{2}\right)=\arcsin(-1)=-\dfrac{\pi}{2}\ne\dfrac{3\pi}{2}$. That means that the property $\arcsin(\sin y)=y$ is not valid for y-values outside the interval $\left[-\dfrac{\pi}{2},\dfrac{\pi}{2}\right]$.

Step 3: The expression $\cos\left(\cos^{-1} 2\pi\right)$ is not defined due to the fact that $\left(\cos^{-1} 2\pi\right)$ is not defined. Remember that the domain of the inverse cosine function is $[-1,1]$. In other words, the value of the cosine is never more than 1, but the value of $2\pi \approx 6.28$ is definitely more than 1.

Solution: You find that $\tan(\arctan(-7)) = -7$, $\arcsin\left(\sin\dfrac{5\pi}{3}\right) = -\dfrac{\pi}{3}$, and $\cos\left(\cos^{-1} 2\pi\right)$ is not defined.

The next problem introduces the method of finding the exact values of functions of inverse functions using right triangles. You learn how to use triangles to convert trigonometric expressions into algebraic expressions.

Sample Problem 5

Find the exact value of $\tan\left(\arccos\dfrac{5}{13}\right)$.

Step 1: You are looking for the tangent of the angle whose cosine is $\dfrac{5}{13}$. Denote an angle whose cosine is $\dfrac{5}{13}$ as v, then you can state that $\angle v = \arccos\dfrac{5}{13}$; therefore, $\cos v = \dfrac{5}{13}$.

Step 2: Because $\cos v$ is positive, $\angle v$ is a first-quadrant angle. You can sketch the right triangle and label $\angle v$ as shown in the following illustration.

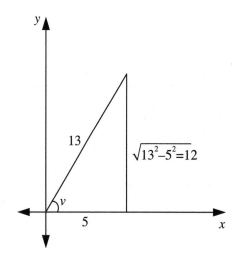

Step 3: Because $\cos = \dfrac{\text{adj.}}{\text{hyp.}} = \dfrac{5}{12}$, you can use the Pythagorean theorem to find the other leg:

$$\text{opp.} = \sqrt{13^2 - 5^2} = \sqrt{169 - 25} = \sqrt{144} = 12$$

The opposite leg is 12.

Step 4: Using trigonometric ratios in right triangles, you can write:

$$\tan v = \frac{\text{opp.}}{\text{adj.}} = \frac{12}{5}$$

Then, by substituting the angle back into the trigonometric expression, you obtain:

$$\tan\left(\arccos\frac{5}{13}\right) = \tan v = \frac{\text{opp.}}{\text{adj.}} = \frac{12}{5}$$

Solution: You find that $\tan\left(\arccos\dfrac{5}{13}\right) = \dfrac{12}{5}$.

In this chapter, you learned about inverse trigonometric functions and saw that they are nothing more than regular trigonometric functions applied "in reverse." Just like regular functions, they can be evaluated and also expressed using right triangles. Last but not least, you have also learned how to evaluate expressions where regular and inverse functions are nested within each other.

Practice Problems

Problem 1: Find the exact value of $\arcsin\dfrac{1}{2}$.

Problem 2: Find the exact value of $\arctan\dfrac{\sqrt{3}}{3}$.

Problem 3: Find the exact value of $\cos(\arccos(-0.1))$.

Problem 4: Find the exact value of $\arcsin(\sin 3\pi)$.

Problem 5: Find the exact value of $\sec\left[\arctan\left(-\dfrac{3}{5}\right)\right]$.

The Least You Need to Know

- The domain of $y = \arcsin x$ is $[-1,1]$, and the range is $-\dfrac{\pi}{2} \leq y \leq \dfrac{\pi}{2}$.

- The domain of $y = \arccos x$ is $[-1,1]$, and the range is $0 \leq y \leq \pi$.

- The domain of $y = \arctan x$ is $-\infty < y < +\infty$, and the range is $-\dfrac{\pi}{2} < y < \dfrac{\pi}{2}$.

- The inverse functions properties for the arcsine are as follows:

  If $-1 \leq x \leq 1$ and $-\dfrac{\pi}{2} \leq y \leq \dfrac{\pi}{2}$, then $\sin(\arcsin x) = x$ and $\arcsin(\sin y) = y$.

- The inverse functions properties for other trigonometric functions are as follows:

  If $-1 \leq x \leq 1$ and $-0 \leq y \leq \pi$, then $\cos(\arccos)x = x$ and $\arccos(\cos y) = y$.

  If $-\dfrac{\pi}{2} < y < \dfrac{\pi}{2}$, then $\tan(\arctan)x = x$ and $\arctan(\tan y) = y$.

Trigonometric Identities and Equations

This part includes all imaginable trigonometric formulas. We start off with reviewing the so-called fundamental trigonometric identities that prepare us for the challenges of the subsequent chapters of this part.

Next, you learn how to verify trigonometric identities and solve trigonometric equations. You will rely heavily on algebra techniques, factorization, collecting like terms, squaring, and the quadratic formula to get to the correct answers. You learn how to obtain general solutions for trigonometric equations using the period of trigonometric functions.

Later, we discuss six categories of formulas or identities that allow you to solve more trigonometric problems. You deal with sums and differences of angles and trigonometric functions, functions' products and powers, and double- and half-angles.

We wrap up this part with some advice on how not to be lost and frustrated by all these formulas, and we suggest some memory aids that can help you memorize these formulas and keep them handy when you need them.

Kaleidoscope of Identities

In This Chapter

- Recalling reciprocal identities
- Listing of cofunction identities
- Modifying basic identities
- Using basic identities to evaluate functions
- Simplifying trigonometric expressions

So far, you have studied basic identities; learned how to solve triangles; explored curves of trigonometric functions; and applied trigonometry to surveying, navigation, and bearing problems. This chapter prepares you to develop additional trigonometric identities and to simplify difficult trigonometric expressions. This chapter also lays the groundwork for solving more complex trigonometric equations.

Many tasks in this chapter and in the following chapters require your algebra skills, such as factoring, performing operations with fractional expressions, rationalizing denominators, and simplifying complex fractions. We guide you through how to tackle these tasks.

This chapter also serves as a springboard to the next two chapters where you move full force into fighting and conquering the toughest identities and equations. Here, you gather your forces and ensure you know how to use them in the most efficient way.

Basic Identities

You might notice that most of the relationships discussed in this chapter have already been mentioned in the preceding chapters. They have been scattered throughout the first half of this book—here, they're listed all in one place, and we explain and expand on some of them in more detail. We try to indicate, whenever possible, how you can apply these "simple" identities in the future to prove the more complex identities awaiting you or to solve trigonometric equations.

Reciprocal Identities

Many of the identities that we talked about are reciprocal (also called quotient) identities. They are the simplest of all identities, but they are handy—you'll resort to their help often. They are listed here.

Quotient Identities

$$\tan \theta = \frac{\sin \theta}{\cos \theta} \qquad \cot \theta = \frac{\cos \theta}{\sin \theta}$$

Reciprocal Identities

$$\sin \theta = \frac{1}{\csc \theta} \qquad \cos \theta = \frac{1}{\sec \theta} \qquad \tan \theta = \frac{1}{\cot \theta}$$

$$\cot \theta = \frac{1}{\tan \theta} \qquad \sec \theta = \frac{1}{\cos \theta} \qquad \csc \theta = \frac{1}{\sin \theta}$$

Let's demonstrate one of the tricks of reciprocal identities. When you multiply the reciprocals together, you get 1:

$$\sin \theta \cdot \csc \theta = 1 \qquad \cos \theta \cdot \sec \theta = 1 \qquad \tan \theta \cdot \cot \theta = 1$$

WORTH KNOWING

Using the graphing utility of your calculator, you can determine whether a given identity is true. Sketch the graphs of each side of the identity and see whether they overlap.

This quality of the reciprocals becomes helpful while solving trigonometric equations. For example, if you multiply each side of the equation by a function's reciprocal, it helps to reduce one side of the equation, or at least some part of it, to 1.

Relationships with Negatives

Recall that in Chapter 10, while discussing even and odd functions, we proved that $\sin(-\theta) = -\sin\theta$ and $\cos(-\theta) = \cos\theta$. These are called negative angle identities. Similar relationships can be established for other trigonometric functions using reciprocal and quotient relationships.

> ### Negative Angle Identities
>
> $$\sin(-\theta) = -\sin\theta \qquad \cos(-\theta) = \cos\theta \qquad \tan(-\theta) = -\tan\theta$$
>
> $$\cot(-\theta) = -\cot\theta \qquad \sec(-\theta) = \sec\theta \qquad \csc(-\theta) = -\csc\theta$$

There are two even (the cosine and secant) and four odd (the sine, cosecant, tangent, and cotangent) trigonometric functions.

Sometimes, these identities are called even-odd identities.

Revisiting Pythagorean Identities

We proved Pythagorean identities in Chapters 4 and 5. Here they are again:

> ### Pythagorean Identities
>
> $$\sin^2\theta + \cos^2\theta = 1 \qquad 1 + \tan^2\theta = \sec^2\theta \qquad 1 + \cot^2\theta = \csc^2\theta$$

Note that the Pythagorean identities with the tangent and cotangent can be rewritten as:

$$\tan^2\theta + 1 = \sec^2\theta \qquad \cot^2\theta + 1 = \csc^2\theta$$

They are still the same thing regardless of the order of the addition on the left side.

The first Pythagorean identity is the most frequently used identity of all—and it is quite powerful, too, because it sets the combination of functions equal to 1.

Let's look at some of the most common uses of the Pythagorean identities. As a warm-up, isolate either the square of the sine or cosine on the left side by subtracting the other term:

$$\sin^2\theta = 1 - \cos^2\theta$$

$$\cos^2\theta = 1 - \sin^2\theta$$

Next, you can modify the previous result further and get the functions by themselves by taking the square roots of both sides of the equations:

$$\sin\theta = \pm\sqrt{1-\cos^2\theta}$$
$$\cos\theta = \pm\sqrt{1-\sin^2\theta}$$

Or you can factor the right sides of the previous set of equations using the formula for the *difference of two squares:*

$$\sin^2\theta = 1-\cos^2\theta = (1+\cos\theta)(1-\cos\theta)$$
$$\cos^2\theta = 1-\sin^2\theta = (1+\sin\theta)(1-\sin\theta)$$

DEFINITION

The **difference of two squares** can be expressed using the following pattern:

$a^2 - b^2 = (a + b)(a - b)$

This last variation of the Pythagorean identity seen previously becomes handy when solving trigonometric equations. Similarly, you can modify the other two Pythagorean identities. Let's show the string of modifications for each of them.

$$\tan^2\theta + 1 = \sec^2\theta \qquad\qquad \cot^2\theta + 1 = \csc^2\theta$$
$$\tan^2\theta = \sec^2\theta - 1 \qquad\qquad \cot^2\theta = \csc^2\theta - 1$$
$$\tan\theta = \pm\sqrt{\sec^2\theta - 1} \qquad\qquad \cot\theta = \pm\sqrt{\csc^2\theta - 1}$$
$$\tan^2\theta = \sec^2\theta - 1 = (\sec\theta + 1)(\sec\theta - 1) \quad \cot^2\theta = \csc^2\theta - 1 = (\csc\theta + 1)(\csc\theta - 1)$$

Cofunction Identities

The sine and cosine are called cofunctions, as are the tangent and cotangent, and the secant and cosecant. There is a special relationship between each pair of a function and its cofunction.

In general, the function of θ = the cofunction of the complement of θ.

In Chapter 4, we explored the cofunction identity between the sine and cosine while studying right triangles. All other similar identities can be derived from this one. The following list is a summary of all cofunction identities.

Cofunction Identities

$$\sin\theta = \cos\left(\frac{\pi}{2} - \theta\right) \qquad \cos\theta = \sin\left(\frac{\pi}{2} - \theta\right)$$

$$\tan\theta = \cot\left(\frac{\pi}{2} - \theta\right) \qquad \cot\theta = \tan\left(\frac{\pi}{2} - \theta\right)$$

$$\sec\theta = \csc\left(\frac{\pi}{2} - \theta\right) \qquad \csc\theta = \sec\left(\frac{\pi}{2} - \theta\right)$$

The basic identities that we just recalled are best known as *fundamental trigonometric identities*. There certainly seem to be a lot of them, and you may wonder whether you need to memorize them all. We suggest that you do. They are indispensable in proving difficult trigonometric identities and solving challenging trigonometric equations. It is helpful to have them handy in your memory so you don't have to flip through the pages of your book whenever you need one or another.

DEFINITION

The **fundamental trigonometric identities** are equalities that involve trigonometric functions and are true for every single value of the occurring variables. They are comprised of reciprocal, quotient, cofunction, even-odd, and Pythagorean identities.

Many Uses of the Fundamental Identities

In this section, you apply the fundamental identities to solve the most common types of problems. One common application is to use them to find values of other trigonometric functions when one or two values are given.

WORTH KNOWING

A quadratic trinomial can be factored into a product of two binomials as:

$3x^2 + 2x - 1 = (3x - 1)(x + 1)$

The goal is to find a combination of factors of x^2 and factors of the free term (-1) such that the inner product and the outer product add to the middle term bx $(2x)$: $(-1) \cdot x + (3x) \cdot 1 = -x + 3x = 2x$.

Sample Problem 1

If $\sec\theta = -\dfrac{5}{3}$ and $\tan\theta > 0$, find the values of all six trigonometric functions.

Step 1: Using a reciprocal identity, you can find the cosine:

$$\cos\theta = \frac{1}{\sec\theta} = \frac{1}{-\dfrac{5}{3}} = -\frac{3}{5}$$

Step 2: Using the Pythagorean identity, you can obtain:

$$\sin^2\theta = 1 - \cos^2\theta = 1 - \left(-\frac{3}{5}\right)^2 = 1 - \frac{9}{25} = \frac{16}{25}$$

Step 3: Because the secant is negative and $\tan\theta > 0$, it follows that angle θ lies in quadrant III. Therefore, you need to choose the negative root to obtain the sine, because the sine is negative for the angles of quadrant III:

$$\sin\theta = -\sqrt{\frac{16}{25}} = -\frac{4}{5}$$

Step 4: Knowing the values for the sine and cosine, you can find the values of all other trigonometric functions:

$$\tan\theta = \frac{\sin\theta}{\cos\theta} = \frac{-\dfrac{4}{5}}{-\dfrac{3}{5}} = \frac{4\cdot 5}{5\cdot 3} = \frac{4}{3}$$

$$\cot\theta = \frac{1}{\tan\theta} = \frac{1}{\dfrac{4}{3}} = \frac{3}{4}$$

$$\csc\theta = \frac{1}{\sin\theta} = \frac{1}{-\dfrac{4}{5}} = -\frac{5}{4}$$

Solution: The six trigonometric functions are $\sin\theta = -\dfrac{4}{5}$, $\cos\theta = -\dfrac{3}{5}$, $\tan\theta = \dfrac{4}{3}$, $\cot\theta = \dfrac{3}{4}$, $\sec\theta = -\dfrac{5}{3}$, and $\csc\theta = -\dfrac{5}{4}$.

Another common use of the fundamental identities is to simplify trigonometric expressions that are given in the form of *quadratic trinomials* and provide the answer as a product of two *binomials*. This tactic is useful when solving trigonometric equations. In this type of problem, you need to use some algebra rules to assist trigonometry.

> **DEFINITION**
>
> A **quadratic trinomial** is an algebraic expression in the following form:
>
> $x^2 + bx + c$
>
> A **binomial** is an algebraic expression that has two terms.

Sample Problem 2

Factor the expression $\sec^2 x - \tan x - 3$.

Step 1: As given, this expression cannot be factored because it contains two types of trigonometric functions. You can use the Pythagorean identity $\tan^2 \theta + 1 = \sec^2 \theta$ to rewrite the expression in terms of the tangent only:

$$\sec^2 x - \tan x - 3 = (1 + \tan^2 x) - \tan x - 3$$

Step 2: Combine like terms:

$$(1 + \tan^2 x) - \tan x - 3 = \tan^2 x - \tan x - 2$$

Step 3: Factor the quadratic trinomial:

$$\tan^2 x - \tan x - 2 = (\tan x - 2)(\tan x + 1)$$

Note that $\tan x$ plays the role of a variable x in the trinomial.

Solution: You find that $\sec^2 x - \tan x - 3 = (\tan x - 2)(\tan x + 1)$.

You can also use the fundamental identities to add or subtract two fractional trigonometric expressions. Using algebra rules, you operate with trigonometric fractions as if they are algebraic fractions.

> **DANGEROUS TURN**
>
> Don't forget to multiply each numerator by a factor that changes the original denominator into the least common denominator (LCD) when adding fractional expressions. For example:
>
> $$\frac{3}{(x+2)} + \frac{4}{(x-5)} = \frac{3(x-5) + 4(x+2)}{(x+2)(x-5)} = \frac{3x - 15 + 4x + 8}{(x+2)(x-5)} = \frac{7x - 7}{(x+2)(x-5)}$$

Sample Problem 3

Perform the indicated addition and simplify the result.

$$\frac{\cos\theta}{1+\sin\theta}+\frac{\sin\theta}{\cos\theta}$$

Step 1: The LCD is $(1+\sin\theta)\cdot\cos\theta$. Rewrite the expression as one fraction with the LCD as a new denominator. You also change the numerators by multiplying them by a corresponding factor:

$$\frac{\cos\theta}{1+\sin\theta}+\frac{\sin\theta}{\cos\theta}=\frac{\cos\theta\cdot\cos\theta+\sin\theta(1+\sin\theta)}{(1+\sin\theta)(\cos\theta)}$$

Step 2: Multiply and distribute in the numerator:

$$\frac{\cos\theta\cdot\cos\theta+\sin\theta(1+\sin\theta)}{(1+\sin\theta)(\cos\theta)}=\frac{\cos^2\theta+\sin\theta+\sin^2\theta}{(1+\sin\theta)(\cos\theta)}$$

Step 3: Rearrange the numerator and use the Pythagorean identity $\sin^2\theta+\cos^2\theta=1$:

$$\frac{\cos^2\theta+\sin\theta+\sin^2\theta}{(1+\sin\theta)(\cos\theta)}=\frac{\cos^2\theta+\sin^2\theta+\sin\theta}{(1+\sin\theta)(\cos\theta)}=\frac{1+\sin\theta}{(1+\sin\theta)(\cos\theta)}$$

Step 4: Cancel common factors and use the reciprocal identity:

$$\frac{1+\sin\theta}{(1+\sin\theta)(\cos\theta)}=\frac{1}{\cos\theta}=\sec\theta$$

Solution: You find that $\dfrac{\cos\theta}{1+\sin\theta}+\dfrac{\sin\theta}{\cos\theta}=\sec\theta$.

The last sample problem in this chapter shows a method of converting a fractional expression into one without fractions. There are two techniques that you should store in your memory for future use: multiplication by a *conjugate*, and separation of fractions. These two techniques will help you solve complex trigonometric problems by eliminating fractions within them and converting them into simpler ones.

DEFINITION

A **conjugate** is a binomial whose middle sign is opposite of another binomial with the same terms. $(3x+5)$ and $(3x-5)$ are conjugates.

Multiplication by a conjugate involves taking a fractional expression and multiplying it by a conjugate so that, at the end, we can apply known trigonometric identities and reduce the fractions out. Separation of fractions utilizes algebra to separate a complex fraction into two or more simpler ones that then can be reduced further using trigonometric identities.

Sample Problem 4

Rewrite $\dfrac{1}{1+\cos\theta}$ so that it is not in fractional form.

Step 1: Multiply both the numerator and the denominator by a conjugate of the denominator:

$$\frac{1}{1+\cos\theta}=\frac{1\cdot(1-\cos\theta)}{(1+\cos\theta)(1-\cos\theta)}=\frac{1-\cos\theta}{1-\cos^2\theta}$$

Note that the denominator contains an algebraic formula for the difference of two squares, where $a=1$ and $b=\cos\theta$.

Step 2: Use the Pythagorean identity for the denominator:

$$\frac{1-\cos\theta}{1-\cos\theta^2\theta}=\frac{1-\cos\theta}{\sin^2\theta}$$

Step 3: Rewrite a fraction as a sum of two separate fractions:

$$\frac{1-\cos\theta}{\sin^2\theta}=\frac{1}{\sin^2\theta}-\frac{\cos\theta}{\sin^2\theta}$$

DANGEROUS TURN

To separate two or more fractions, separate the numerator only, never separate the denominator. For example,

$$\frac{x-2y+c}{3+b}=\frac{x-2y}{3+b}+\frac{c}{3+b}\quad\text{or}\quad\frac{x}{3+b}-\frac{2y}{3+b}+\frac{c}{3+b}\quad\text{is correct, but}$$

$$\frac{x-2y+c}{3+b}=\frac{x-2y}{3}+\frac{c}{b}\quad\text{is incorrect.}$$

Step 4: As the last step, rewrite the second fraction as a product of two fractions and use reciprocal identities for each fraction:

$$\frac{1}{\sin^2\theta}-\frac{\cos\theta}{\sin^2\theta}=\frac{1}{\sin^2\theta}-\frac{\cos\theta}{\sin\theta}\cdot\frac{1}{\sin\theta}=\csc^2\theta-\cot\theta\csc\theta$$

Solution: You find that $\dfrac{1}{1+\cos\theta}=\csc^2\theta-\cot\theta\cdot\csc\theta$.

Sample problems of this chapter illustrate several important applications of the fundamental trigonometric identities. We hope that they warm you up for the challenges of the next chapter, where you deal with new, and more complex, trigonometric identities.

Practice Problems

Problem 1: Use the fundamental identities to evaluate the other trigonometric functions if $\tan\theta = 2$ and $\sin\theta < 0$.

Problem 2: Factor the expression and use fundamental identities to simplify: $\cos^2\theta\csc^2\theta - \cos^2\theta$.

Problem 3: Factor the expression and use fundamental identities to simplify: $\cos^4\theta - \sin^4\theta$.

Problem 4: Subtract and simplify: $\cot\theta - \dfrac{\csc^2\theta}{\cot\theta}$.

Problem 5: Rewrite the expression so that it is not in fractional form: $\dfrac{7}{\sec\theta - \tan\theta}$.

The Least You Need to Know

- Trigonometric Pythagorean identities are among the most frequently used basic identities.
- Reciprocal identities help to write trigonometric expressions in terms of the three main trig functions.
- Basic identities help to evaluate trigonometric functions and simplify trigonometric expressions.
- Basic identities help to develop additional trigonometric identities and solve trigonometric equations.

Verifying Trigonometric Identities

Chapter 15

In This Chapter

- General suggestions on proving trig identities
- Looking at the arsenal of the methods of proof
- Dealing with fractions and radicals
- Using algebra rules to help with identities

In this chapter, you learn to prove or verify trigonometric identities. Proving an identity involves taking commonly known facts as a foundation and then performing mathematical operations, the result of which is an equation that represents the identity. The key to success in this endeavor is your ability to use the fundamental identities and the algebra rules to alter or rewrite trigonometric expressions.

Before going on, let's review the distinctions among expressions, equations, and identities. Expressions don't have any equal signs and thus they don't have left or right sides. For example, $\sin\theta + \cos\theta$ is an expression. When simplifying expressions, you use the equal sign only to indicate that the original expression is equivalent to the modified expression. An equation is a mathematical statement containing an equal sign that is true only for a specific set of values. Therefore, an equation can be untrue for some arbitrary values. For example, the equation $\sin x = 1$ is only true for $x = \dfrac{\pi}{2} \pm 2n\pi$. To the contrary, an identity is true for all real values in the domain of the variable. For example, the Pythagorean identity $\sin^2\theta + \cos^2\theta = 1$ is true for all real numbers.

Important Guidelines

Unfortunately, there are no well-defined rules or procedures to follow in verifying trigonometric identities. The best advice is to learn by practice and prove many of them in order to gain experience. However, here are some general guidelines:

- Don't just stare at the problem in frustration—try to tackle it one way or another. You might fail with the first (and the second) try, but eventually something will work.

- Work with one side of the equation at a time. Start with the more complicated side.

- Apply algebraic techniques: factor an expression, add fractions, factor out a monomial (i.e., a single term), square a binomial, etc.

- Look for opportunities to use the fundamental identities.

With that being said, let's start working with the trigonometric identities.

Different Tactics for Verifying Identities

There are many tactics and methods for proving the trigonometric identities. We present some of the most common and often used ones. With each identity, you need to inspect it first and then decide which method to choose.

WORD OF ADVICE

There can be more than one way to verify a trigonometric identity. Your method might be different from the one we use. This provides a chance to be creative and even establish your favorite tactics for proving. However, it's best to be efficient and prove identities the quickest way.

For each method or tactic discussed in this chapter, we provide some rationale and justify why we decided to use this method versus another.

Working with the More Complicated Side

The first tactic involves inspecting the identity and deciding which side looks more complicated. Once you've decided, start working with the complicated side.

Sample Problem 1

Verify the identity $\dfrac{\csc^2\theta - 1}{\csc^2\theta} = \cos^2\theta$.

Step 1: Start with the left side, because it is the more complicated one, and use the Pythagorean identity for the cosecant in the numerator:

$$\frac{\csc^2\theta - 1}{\csc^2\theta} = \frac{\left(\cot^2\theta + 1\right) - 1}{\csc^2\theta}$$

Step 2: Simplify the numerator and use the reciprocal identity:

$$\frac{\cot^2\theta + 1 - 1}{\csc^2\theta} = \frac{\cot^2\theta}{\csc^2\theta} = \cot^2\theta \cdot \sin^2\theta$$

Step 3: Use the quotient identity for the cotangent and simplify:

$$\cot^2\theta \cdot \sin^2\theta = \frac{\cos^2\theta}{\sin^2\theta} \cdot \sin^2\theta = \cos^2\theta$$

Solution: You verify that $\dfrac{\csc^2\theta - 1}{\csc^2\theta} = \cos^2\theta$.

There is another way to do sample problem 1 if you remember that it is sometimes helpful to separate a fraction into two parts. Let's discuss this method.

Step 1: Separate the fraction into two parts:

$$\frac{\csc^2\theta - 1}{\csc^2\theta} = \frac{\csc^2\theta}{\csc^2\theta} - \frac{1}{\csc^2\theta}$$

Note that the minus sign that was in front of 1 on the left side is now in front of the second fraction.

Step 2: Simplify the first fraction and use the reciprocal identity for the second one:

$$\frac{\csc^2\theta}{\csc^2\theta} - \frac{1}{\csc^2\theta} = 1 - \sin^2\theta$$

Step 3: Use the Pythagorean identity:

$$1 - \sin^2\theta = \cos^2\theta$$

Solution: You verify that $\dfrac{\csc^2\theta - 1}{\csc^2\theta} = \cos^2\theta$.

In the next problem, you keep working with the more complicated side and illustrate that it is often more useful to use identities before multiplying the terms.

Sample Problem 2

Verify the identity $\left(\cos^2\theta - 1\right)\left(\tan^2\theta + 1\right) = -\tan^2\theta$.

Step 1: Work with the left side as the more complicated one. Simplify in both parentheses before multiplying. The second parentheses contains the tangent Pythagorean identity, and the first parentheses contains the modified Pythagorean identity. Therefore:

$$\left(\cos^2\theta - 1\right)\left(\tan^2\theta + 1\right) = \left(-\sin^2\theta\right)\left(\sec^2\theta\right)$$

> **WORD OF ADVICE**
>
> The Pythagorean identity can be modified as follows:
>
> $\sin^2\theta + \cos^2\theta = 1 \;\rightarrow\; \sin^2\theta - 1 = -\cos^2\theta \;\rightarrow\; \cos^2\theta - 1 = -\sin^2\theta$

Step 2: Use the reciprocal identity for the secant:

$$\left(-\sin^2\theta\right)\left(\sec^2\theta\right) = \left(-\sin^2\theta\right)\left(\frac{1}{\cos^2\theta}\right) = -\frac{\sin^2\theta}{\cos^2\theta}$$

Step 3: Use the rules of exponents and the tangent quotient identity:

$$-\frac{\sin^2\theta}{\cos^2\theta} = -\left(\frac{\sin\theta}{\cos\theta}\right)^2 = -\tan^2\theta$$

Solution: You verify that $\left(\cos^2\theta - 1\right)\left(\tan^2\theta + 1\right) = -\tan^2\theta$.

Combining Fractions Before Using Identities

If one of the sides of a given identity includes fractions, it is always helpful to combine fractions first before using identities. Fractions, especially trigonometric fractions, often cause fear and discomfort because many people don't feel skilled enough to operate even with simple fractions, let alone trigonometric ones. Still, you'll be pretty safe if you remember just one simple thing when operating with fractions: the least common denominator (LCD). In most identity proving problems, the denominators of two fractions don't have anything in common, so you just need to multiply the denominators of two fractions to get the LCD. Don't forget to change the numerators as well by multiplying them by the other fraction's denominator. Our next problem illustrates this method.

Sample Problem 3

Verify the identity $\dfrac{1}{1-\cos\theta}+\dfrac{1}{1+\cos\theta}=2\csc^2\theta$.

Step 1: Start working with the left side because it is more complicated. Because the denominators of both fractions are different (they don't have anything in common), the LCD is the product of these two denominators. You change the numerators accordingly as well by multiplying each one by the other fraction's denominator:

$$\frac{1}{1-\cos\theta}+\frac{1}{1+\cos\theta}=\frac{1(1+\cos\theta)}{(1-\cos\theta)(1+\cos\theta)}+\frac{1(1-\cos\theta)}{(1-\cos\theta)(1+\cos\theta)}=\frac{1+\cos\theta+1-\cos\theta}{(1-\cos\theta)(1+\cos\theta)}$$

Step 2: Simplify the numerator, and then use the formula for the difference of two squares that was discussed in Chapter 14 for the denominator ($a = 1$, $b = \cos\theta$):

$$\frac{1+\cos\theta+1-\cos\theta}{(1-\cos\theta)(1+\cos\theta)}=\frac{2}{1-\cos^2\theta}$$

Step 3: Use the Pythagorean and reciprocal identities in the denominator:

$$\frac{2}{1-\cos^2\theta}=\frac{2}{\sin^2\theta}=2\csc^2\theta$$

Solution: You verify that $\dfrac{1}{1-\cos\theta}+\dfrac{1}{1+\cos\theta}=2\csc^2\theta$.

Converting to Sines and Cosines

Sometimes when you look at the problem, there is no visible opportunity to use the Pythagorean identities and no fractions to add. However, you might notice that the identity contains a hodgepodge of trigonometric functions. This is a hint to convert everything to sines and cosines.

Sample Problem 4

Verify the identity $\sec\theta-\sin\theta\tan\theta=\cos\theta$.

Step 1: Express each function on the left in terms of the sine and cosine:

$$\sec\theta-\sin\theta\tan\theta=\frac{1}{\cos\theta}-\sin\theta\cdot\frac{\sin\theta}{\cos\theta}=$$
$$=\frac{1}{\cos\theta}-\frac{\sin^2\theta}{\cos\theta}$$

Step 2: Combine two fractions:

$$\frac{1}{\cos\theta} - \frac{\sin^2\theta}{\cos\theta} = \frac{1-\sin^2\theta}{\cos\theta}$$

Step 3: Use the Pythagorean identity in the numerator and then cancel the common factors:

$$\frac{1-\sin^2\theta}{\cos\theta} = \frac{\cos^2\theta}{\cos\theta} = \cos\theta$$

Solution: You verify that $\sec\theta - \sin\theta\tan\theta = \cos\theta$.

Let's do one more problem of this type.

Sample Problem 5

Verify the identity $\dfrac{\tan\theta + \cot\theta}{\sec^2\theta} = \cot\theta$.

Step 1: Express each function on the left in terms of the sine and cosine:

$$\frac{\tan\theta + \cot\theta}{\sec^2\theta} = \frac{\dfrac{\sin\theta}{\cos\theta} + \dfrac{\cos\theta}{\sin\theta}}{\dfrac{1}{\cos^2\theta}}$$

Step 2: You obtain a monstrous-looking complex fraction. Don't let it scare you; continue working on the numerator. Add two fractions there and leave the denominator unchanged:

$$\frac{\dfrac{\sin\theta}{\cos\theta} + \dfrac{\cos\theta}{\sin\theta}}{\dfrac{1}{\cos^2\theta}} = \frac{\dfrac{\sin\theta\sin\theta + \cos\theta\cos\theta}{\cos\theta\sin\theta}}{\dfrac{1}{\cos^2\theta}} =$$

$$= \frac{\dfrac{\sin^2\theta + \cos^2\theta}{\cos\theta\sin\theta}}{\dfrac{1}{\cos^2\theta}}$$

Step 3: Apply the Pythagorean identity in the numerator:

$$\frac{\dfrac{\sin^2\theta+\cos^2\theta}{\cos\theta\sin\theta}}{\dfrac{1}{\cos^2\theta}}=\frac{\dfrac{1}{\cos\theta\sin\theta}}{\dfrac{1}{\cos^2\theta}}$$

Step 4: Deal with the complex fraction first (remember, $\dfrac{\dfrac{a}{b}}{\dfrac{c}{d}}=\dfrac{a\cdot d}{b\cdot c}$), and then simplify:

$$\frac{\dfrac{1}{\cos\theta\sin\theta}}{\dfrac{1}{\cos^2\theta}}=\frac{\cos^2\theta}{\cos\theta\sin\theta}$$

$$\frac{\cos^2\theta}{\cos\theta\sin\theta}=\frac{\cos\theta}{\sin\theta}=\cot\theta$$

Solution: You verify that $\dfrac{\tan\theta+\cot\theta}{\sec^2\theta}=\cot\theta.$

Working with Each Side Separately

So far, you have been working with one (more complicated) side of the identity and converting it into the form given on the other side. Sometimes it is more practical to work with each side separately and obtain one common form that is equivalent for both sides. You can use this approach whenever it is difficult to decide which side is more complicated. The next problem shows how to implement this approach.

Sample Problem 6

Verify the identity $\dfrac{\tan^2\theta}{1+\sec\theta}=\dfrac{1-\cos\theta}{\cos\theta}.$

Step 1: Work with the left side first and apply the Pythagorean identity in the numerator:

$$\frac{\tan^2\theta}{1+\sec\theta}=\frac{\sec^2\theta-1}{1+\sec\theta}$$

Step 2: Factor the difference of two squares in the numerator and reduce:

$$\frac{\sec^2\theta-1}{1+\sec\theta}=\frac{(\sec\theta+1)(\sec\theta-1)}{1+\sec\theta}=\sec\theta-1$$

Step 3: Now, work with the right side and separate the fraction into two:

$$\frac{1-\cos\theta}{\cos\theta}=\frac{1}{\cos\theta}-\frac{\cos\theta}{\cos\theta}$$

Step 4: Use the reciprocal identity for the first term and reduce the second fraction:

$$\frac{1}{\cos\theta} - \frac{\cos\theta}{\cos\theta} = \sec\theta - 1$$

You have proven that both sides are equal to $\sec\theta - 1$.

Solution: You verify that $\dfrac{\tan^2\theta}{1+\sec\theta} = \dfrac{1-\cos\theta}{\cos\theta}$.

Other Useful Techniques

Sometimes the method of multiplying by a conjugate helps with some identities. A hint to apply this method is the presence of expressions such as $1+\sin\theta$ or $\cos\theta - 1$. But why does multiplication by a conjugate help? It helps because the result is usually the Pythagorean identity, as in the following example.

Suppose you have an expression $1+\sin\theta$ in the denominator. Let's see what happens when you multiply it by its conjugate:

$1+\sin\theta$	Multiply by a conjugate.
$(1+\sin\theta)(1-\sin\theta)$	Use the difference of two squares formula.
$1-\sin^2\theta$	Finally, use the Pythagorean identity.
$\cos^2\theta$	The result is obtained.

Multiplication by a conjugate enables us to substitute a given denominator with the simpler one that, in turn, can be cancelled out with factors in the numerator or will enable a separation of fractions as the next sample problem illustrates.

When you use this technique, multiply by a conjugate both the numerator and the denominator.

Sample Problem 7

Verify the identity $\csc\theta + \cot\theta = \dfrac{\sin\theta}{1-\cos\theta}$.

Step 1: Work with the right side and multiply it by a conjugate of the denominator:

$$\frac{\sin\theta}{1-\cos\theta} = \frac{\sin\theta(1+\cos\theta)}{(1-\cos\theta)(1+\cos\theta)}$$

Step 2: Distribute the expression in the numerator and use the difference of two squares formula in the denominator:

$$\frac{\sin\theta(1+\cos\theta)}{(1-\cos\theta)(1+\cos\theta)} = \frac{\sin\theta - \sin\theta\cos\theta}{1-\cos^2\theta}$$

Step 3: Use the Pythagorean identity in the denominator and then separate fractions:

$$\frac{\sin\theta - \sin\theta\cos\theta}{1-\cos^2\theta} = \frac{\sin\theta - \sin\theta\cos\theta}{\sin^2\theta} = \frac{\sin\theta}{\sin^2\theta} - \frac{\sin\theta\cos\theta}{\sin^2\theta}$$

Step 4: Reduce and use the reciprocal and quotient identities:

$$\frac{\sin\theta}{\sin^2\theta} - \frac{\sin\theta\cos\theta}{\sin^2\theta} = \frac{1}{\sin\theta} - \frac{\cos\theta}{\sin\theta} = \csc\theta - \cot\theta.$$

Solution: You verify that $\csc\theta + \cot\theta = \dfrac{\sin\theta}{1-\cos\theta}$.

The last technique we discuss in this chapter is the method of squaring both sides of the identity. You resort to this drastic method only if one of the sides has a radical. The next example illustrates this method.

Sample Problem 8

Verify the identity $\dfrac{1-\tan\theta}{\sec\theta} = \sqrt{1-2\sin\theta\cos\theta}$.

Step 1: You work on both sides simultaneously and square both sides, because the right side has a radical:

$$\left(\frac{1-\tan\theta}{\sec\theta}\right)^2 = \left(\sqrt{1-2\sin\theta\cos\theta}\right)^2$$

Step 2: When squaring the left side, recall that the rules of exponents dictate to square both the numerator and the denominator:

$$\frac{(1-\tan\theta)^2}{\sec^2\theta} = 1-2\sin\theta\cos\theta$$

Watch out for the correct application of the multiplication pattern while squaring the binomial in the numerator on the left. Note that you keep the right side unchanged until Step 6:

$$\frac{1-2\tan\theta+\tan^2\theta}{\sec^2\theta} = 1-2\sin\theta\cos\theta$$

> **WORD OF ADVICE**
>
> When squaring a binomial, use the following patterns: $(a + b)^2 = a^2 + 2ab + b^2$ and $(a - b)^2 = a^2 - 2ab + b^2$.

Step 3: Rearrange the terms in the numerator and apply the Pythagorean identity for the tangent afterward:

$$\frac{1 + \tan^2 \theta - 2\tan \theta}{\sec^2 \theta} = 1 - 2\sin \theta \cos \theta$$

$$\frac{\sec^2 \theta - 2\tan \theta}{\sec^2 \theta} = 1 - 2\sin \theta \cos \theta$$

Step 4: Separate the fractions on the left and simplify the first term on the left side:

$$\frac{\sec^2 \theta}{\sec^2 \theta} - \frac{2\tan \theta}{\sec^2 \theta} = 1 - 2\sin \theta \cos \theta$$

$$1 - \frac{2\tan \theta}{\sec^2 \theta} = 1 - 2\sin \theta \cos \theta$$

Step 5: Rewrite the second term on the left by using the quotient identity for the tangent and the reciprocal identity for the secant:

$$1 - 2 \cdot \frac{\dfrac{\sin \theta}{\cos \theta}}{\dfrac{1}{\cos^2 \theta}} = 1 - 2\sin \theta \cos \theta$$

Step 6: Simplify the complex fraction and cancel common factors on the left side:

$$1 - \frac{2 \cdot \sin \theta \cos^2 \theta}{1 \cdot \cos \theta} = 1 - 2\sin \theta \cos \theta$$

$$1 - \frac{2 \cdot \sin \theta \cos^{\cancel{2}} \theta}{\cancel{\cos \theta}} = 1 - 2\sin \theta \cos \theta$$

$$1 - 2\sin \theta \cos \theta = 1 - 2\sin \theta \cos \theta$$

Both sides are equal, so you proved the identity.

Solution: You verify that $\dfrac{1 - \tan \theta}{\sec \theta} = \sqrt{1 - 2\sin \theta \cos \theta}$.

We discussed the various methods of proving trigonometric identities. But why do you need to prove them? Did the question of why you need to prove something that is already proven haunt you? If this was the case, you can find relief in realizing that, by proving identities, you just did the major groundwork and training to prepare yourself to solve the trig equations, which will be the topic of the next chapter.

Practice Problems

Problem 1: Verify the identity $\dfrac{\sin\theta\tan\theta}{1-\cos\theta}-1=\sec\theta$.

Problem 2: Verify the identity $\dfrac{\csc\theta-1}{1-\sin\theta}=\csc\theta$.

Problem 3: Verify the identity $\dfrac{1+\sec(-\theta)}{\sin(-\theta)+\tan(-\theta)}=-\csc\theta$.

Problem 4: Verify the identity $\sqrt{\dfrac{1+\cos\theta}{1-\cos\theta}}=\dfrac{1+\cos\theta}{|\sin\theta|}$.

Problem 5: Verify the identity $\dfrac{\cot\theta}{\csc\theta-1}=\dfrac{\csc\theta+1}{\cot\theta}$.

The Least You Need to Know

- Inspect the identity closely before choosing the appropriate method of proof.
- In proving an identity, it is often helpful to start with the more complicated side.
- It is often helpful to add fractions before using the fundamental identities.
- Converting everything to sines and cosines helps when other options are not so obvious.
- When both of the sides are complicated, work with both of them simultaneously.
- Squaring both sides helps when one of the sides is a radical.

Solving Trigonometric Equations

In This Chapter

- Finding solutions within limits as well as general solutions
- Using algebra techniques to solve trig equations
- Exploring trigonometric equations in quadratic form
- Dealing with multiple-angle equations

In this chapter, we make the switch from verifying identities to solving trigonometric equations. To see the difference, consider two equations: $1 + \tan^2 \theta = \sec^2 \theta$ and $\cos \theta = 1$. The first equation is an identity because it is true for all angle values. The second equation is true only for some of the angle's values.

What It Means to Solve a Trigonometric Equation

To solve a trigonometric equation means to find the angle values for which the equation is true. When you find these values, you have solved the equation.

Let's consider a simple trigonometric equation and try to solve it:

$$2\sin x - 1 = 0$$

Add 1 to both sides and divide each side by 2:

$$2\sin x = 1$$
$$\sin x = \frac{1}{2}$$

Now, look for a special angle, if you can, with a sine equal to $\frac{1}{2}$. You know that this is the angle $x = \frac{\pi}{6}$. From studying the unit circle, you also know that there is another angle like that in the second quadrant, $x = \frac{5\pi}{6}$, that has the same value for the sine. Thus you have two solutions in the main interval $[0, 2\pi]$. Because the sine has a period of 2π, there are infinitely many other solutions outside the main interval because new solutions are obtained with each new revolution around the unit circle. You can write all these solutions as $x = \frac{\pi}{6} + 2n\pi$ and $x = \frac{5\pi}{6} + 2n\pi$, where n is an integer. These are called *the general form of the solution.*

> **DEFINITION**
>
> Trigonometric functions are periodic functions; thus there are infinitely many solutions for trigonometric equations. For example, the equation $2\sin x - 1 = 0$ has two solutions for the $0 \le x \le 2\pi$ interval, which are $x = \frac{\pi}{6}$ and $x = \frac{5\pi}{6}$. Solutions that include all possible angles are called **the general form of the solution.** For the given equation, they are $x = \frac{\pi}{6} + 2n\pi$ and $x = \frac{5\pi}{6} + 2n\pi$.

You approach trigonometric equations the same way as you deal with regular algebraic equations. The only difference is that instead of familiar x's and y's, you have trigonometric functions.

Solving Techniques

Because solving trigonometric equations is similar to solving algebraic equations, the techniques are similar as well. In this section, we consider some of the most popular ones. The first technique involves collecting like terms.

Sample Problem 1

Solve $\cos x + \sqrt{2} = -\cos x$.

Step 1: Add $\cos x$ to both sides and subtract $\sqrt{2}$ from both sides:

$$\cos x + \sqrt{2} = -\cos x$$
$$\cos x + \cos x = -\sqrt{2}$$

Step 2: Collect like terms and divide both sides by 2:

$$2\cos x = -\sqrt{2}$$

$$\cos x = -\frac{\sqrt{2}}{2}$$

Step 3: Find the solutions within the main interval $0 \leq x \leq 2\pi$. These are $x = \frac{3\pi}{4}$ and $x = \frac{5\pi}{4}$.

Step 4: Next, add $2n\pi$ (because the period for the cosine is 2π) to each of these solutions to obtain the general form:

$$x = \frac{3\pi}{4} + 2n\pi \text{ and } x = \frac{5\pi}{4} + 2n\pi$$

Solution: You find that $x = \frac{3\pi}{4} + 2n\pi$ and $x = \frac{5\pi}{4} + 2n\pi$.

Extracting square roots whenever possible is another common technique in solving equations. Usually, trigonometric equations are pretty short (in terms of having not too many terms) and you can extract the square root after some simple algebraic operations.

Sample Problem 2

Solve the equation $3\tan^2 x - 1 = 0$.

Step 1: Add 1 to both sides:

$$3\tan^2 x - 1 = 0$$

$$3\tan^2 x = 1$$

Step 2: Divide both sides by 3:

$$\tan^2 x = \frac{1}{3}$$

Step 3: Extract square roots:

$$\tan x = \pm\frac{1}{\sqrt{3}} = \pm\frac{\sqrt{3}}{3}$$

Step 4: Because the tangent has a period of π, find the solutions within the interval $0 \leq x \leq \pi$. These are $x = \frac{\pi}{6}$ (for a positive value) and $x = \frac{5\pi}{6}$ (for a negative value).

Step 5: Next, add $n\pi$ (because the period for the tangent is π) to each of these solutions to obtain the general form:

$$x = \frac{\pi}{6} + n\pi \text{ and } x = \frac{5\pi}{6} + n\pi$$

Solution: You find that $x = \frac{\pi}{6} + n\pi$ and $x = \frac{5\pi}{6} + n\pi$.

> **WORD OF ADVICE**
>
> To find solutions within the main interval, refer to the illustration of the unit circle with coordinates from Chapter 9 for help. The values of the trigonometric functions for special angles will help as well.

When two or more functions occur in the same equation, separate it by factoring. This is the same type of factoring you use when solving regular equations. The most common factoring patterns are the greatest common factor, the difference of two squares, and factoring trinomials. Some of these were discussed when we verified trigonometric identities in Chapters 4 and 5.

Sample Problem 3

Solve the equation $\tan x \sin^2 x = 2\tan x$.

Step 1: Subtract $2\tan x$ from both sides:

$$\tan x \sin^2 x = 2\tan x$$

$$\tan x \sin^2 x - 2\tan x = 0$$

Step 2: Factor the left side:

$$\tan x (\sin^2 x - 2) = 0$$

Step 3: Set each factor to be equal to zero:

$$\tan x = 0 \text{ and } \sin^2 x - 2 = 0$$

Setting each factor equal to zero is a helpful technique to use because if one factor were to equal zero, the whole equation would, too. Therefore, a solution that makes one factor equal to zero would be a solution for the whole equation.

Step 4: Obtain the solution for the second equation. Add 2 to both sides of the second equation and extract the square root:

$$\sin^2 x - 2 = 0$$

$$\sin^2 x = 2$$

$$\sin x = \pm\sqrt{2}$$

No solutions can be obtained from this second equation because $\pm\sqrt{2}$, which is either −1.41 or 1.41, is outside of the range of the sine function. You can recall that the range of the sine function is from −1 to 1.

If a part of an equation has no solution, this simply means that the original equation has a smaller number of solutions than it otherwise would. There could be cases where no factored parts of an equation would have a solution, in which case the original equations would have no solutions as well. In this problem, though, we still have the first equation left to work with.

Step 5: Find solutions for the first equation:

$$\tan x = 0$$

$$x = 0$$

The general form is $x = n\pi$.

Solution: You find that $x = n\pi$.

Equations of Quadratic Type

Many trigonometric equations are of a quadratic type such as $\sin^2 x - 3\sin x - 1 = 0$. To solve such an equation, factor it as if it were a trinomial $x^2 - 3x - 1 = 0$, where x stands for the sine. Let's see how to do this.

Sample Problem 4

Solve the equation $2\cos^2 x + 3\sin x - 3 = 0$.

Step 1: You have two types of functions in the equation. This is equivalent to factoring a trinomial with both x's and y's as variables. Obtain an equation with one type of trig function by using the Pythagorean identity:

$$2\cos^2 x + 3\sin x - 3 = 0$$

$$2(1 - \sin^2 x) + 3\sin x - 3 = 0$$

Step 2: Distribute on the left side, collect like terms, and multiply by –1 both sides:

$$2 - 2\sin^2x + 3\sin x - 3 = 0$$

$$-2\sin^2x + 3\sin x - 1 = 0$$

$$2\sin^2x - 3\sin x + 1 = 0$$

Step 3: Factor using a factoring a trinomial pattern:

$$(2\sin x - 1)(\sin x - 1) = 0$$

Step 4: Set each factor equal to zero and find the solutions in the interval $[0, 2\pi]$:

For $\sin x - 1 = 0$ or $\sin x = 1$, the solution is $x = \dfrac{\pi}{2}$.

For $2\sin x - 1 = 0$ or $\sin x = \dfrac{1}{2}$, there are two solutions, $x = \dfrac{\pi}{6}$ and $x = \dfrac{5\pi}{6}$.

Step 5: Find solutions in general form:

$$x = \frac{\pi}{2} + 2n\pi, \ x = \frac{\pi}{6} + 2n\pi, \text{ and } x = \frac{5\pi}{6} + 2n\pi$$

Solution: You find that $x = \dfrac{\pi}{2} + 2n\pi$, $x = \dfrac{\pi}{6} + 2n\pi$, and $x = \dfrac{5\pi}{6} + 2n\pi$.

Sometimes you need to use the inverse functions to obtain solutions to the equation, as featured in the next example. This is the case when the obtained value of the trigonometric function does not correspond to any special angle.

Sample Problem 5

Solve $\tan^2x - 2\tan x = 3$.

Step 1: Subtract 3 from both sides and factor the left side:

$$\tan^2x - 2\tan x = 3$$

$$\tan^2x - 2\tan x - 3 = 0$$

$$(\tan x - 3)(\tan x + 1) = 0$$

Step 2: Setting each factor equal to zero produces two solutions on the interval $\left[-\dfrac{\pi}{2}, \dfrac{\pi}{2}\right]$:

$$\tan x = 3 \rightarrow x = \arctan 3$$

$$\tan x = -1 \rightarrow x = -\frac{\pi}{4}$$

Because you don't know any special angle with a tangent equal to 3, you need to use an inverse function to write the solution. Another option, given a scientific or graphing calculator, is to compute the answer numerically.

Step 3: Add multiples of π (the period of the tangent) to obtain the general solutions:

$$x = \arctan 3 + n\pi \text{ (can be alternatively written as } x \approx 1.249 + n\pi \text{) and}$$
$$x = -\frac{\pi}{4} + n\pi$$

Solution: You find that $x = \arctan 3 + n\pi$ (or $x \approx 1.249 + n\pi$) and $x = -\frac{\pi}{4} + n\pi$.

When it is impossible to factor out the trigonometric equation of quadratic type, use the *quadratic formula* to obtain the solutions. But how do you learn that you cannot factor an equation? Usually, you learn by trying. Try to factor an equation, and if you fail, resort to using the quadratic formula. Another way to decide whether an equation can be factored is by utilizing the quadratic formula and looking at the *discriminant*.

> **DEFINITION**
>
> The solutions of the quadratic equations $ax^2 + bx + c = 0$ are obtained by
>
> utilizing the **quadratic formula** $x = \dfrac{-b \pm \sqrt{b^2 - 4ac}}{2a}$.
>
> The expression $b^2 - 4ac$ is called a **discriminant.**

If the discriminant is positive, then there are two solutions; if it is equal to zero, then there is only one solution. If the discriminant is negative, then there is no solution to the equation.

Also, if the discriminant is not a perfect square, then it is impossible to factor the trinomial; this means you definitely need to use the quadratic formula to solve the equation.

Sample Problem 6

Find solutions of $\sin^2 x - 3\sin x - 2 = 0$ in the interval $[0, 2\pi]$.

Step 1: Calculate the discriminant to see whether you can factor the left side:

$$b^2 - 4ac = (-3)^2 - 4(1)(-2) = 9 + 8 = 17$$

Because it is not a perfect square, you cannot factor the left side and need to use the quadratic formula.

Step 2: Use the quadratic formula:

$$\sin x = \frac{-(-3) \pm \sqrt{(-3)^2 - 4(1)(-2)}}{2(1)} = \frac{3 \pm \sqrt{17}}{2}$$

$\sin x \approx 3.5616$ or $\sin x \approx -0.5616$

Because the sine cannot be more than 1, discard the first solution.

Step 3: For the second solution, use the inverse function to find the angle in radians:

$$x \approx \arcsin(-0.5616) \approx -0.5963 \approx -0.60$$

Step 4: Note that this solution is not in the interval $[0, 2\pi]$ because it is a negative angle (–0.60 radians). Coterminal positive angle for a –0.60 radian angle is $x \approx 2\pi - 0.60 \approx 5.68$. Another angle that has the same (negative) sine value is the angle in the third quadrant, $x \approx \pi + 0.60 \approx 3.74$.

Solution: You find that $x \approx 5.68$ and $x \approx 3.74$.

When Multiple Angles Are Involved

Multiple-angle equations are those in which the angle measure is some multiple of a variable. For example, instead of $\sin x$, you might have $\sin 3x$ or $\cos 5x$. Functions involving multiple angles require an increased number of solutions. The larger the multiplier of the angle, the more solutions you have.

Sample Problem 7

Solve $2\sin 3x - 1 = 0$.

Step 1: Add 1 to both sides and then divide each side by 2:

$2\sin 3x - 1 = 0$

$2\sin 3x = 1$

$\sin 3x = \dfrac{1}{2}$

Step 2: In the interval $[0, 2\pi]$, there are two angles with sine equal to $\dfrac{1}{2}$:

$$3x = \frac{\pi}{6} \text{ and } 3x = \frac{5\pi}{6}$$

Then, general solutions are

$$3x = \frac{\pi}{6} + 2n\pi \text{ and } 3x = \frac{5\pi}{6} + 2n\pi$$

Step 3: Find x by dividing by 3:

$$x = \frac{\pi}{18} + \frac{2n\pi}{3} \text{ and } x = \frac{5\pi}{18} + \frac{2n\pi}{3}$$

Solution: You find that $x = \frac{\pi}{18} + \frac{2n\pi}{3}$ and $x = \frac{5\pi}{18} + \frac{2n\pi}{3}$.

Note that in the previous problem, because of the multiple-angle involvement, you have six solutions within the interval $[0, 2\pi]$: $\frac{\pi}{18}$, $\frac{13\pi}{18}$, $\frac{25\pi}{18}$, $\frac{5\pi}{18}$, $\frac{17\pi}{18}$, and $\frac{29\pi}{18}$. You obtain them by assigning values of 1 and 2 for the integer n in the general solution. The larger values of n produce solutions outside the given interval.

Your last problem involves an angle that has a fractional multiplier.

Sample Problem 8

Solve $6\tan\left(\frac{x}{2}\right) + 6 = 0$.

Step 1: Subtract 6 from both sides and then divide each side by 6:

$$6\tan\left(\frac{x}{2}\right) + 6 = 0$$

$$6\tan\left(\frac{x}{2}\right) = -6$$

$$\tan\frac{x}{2} = -1$$

Step 2: In the interval $[0, \pi]$, there is one angle with a tangent is equal to -1:

$$\frac{x}{2} = \frac{3\pi}{4}$$

Then, general solution is:

$$\frac{x}{2} = \frac{3\pi}{4} + n\pi$$

Step 3: Find x by multiplying by 2:

$$x = \frac{3\pi}{2} + 2n\pi$$

Solution: You find that $x = \frac{3\pi}{2} + 2n\pi$.

We consider many methods for solving trig equations, but certainly we could not cover all of the possible cases. Practice problems provide some additional techniques. For a wrap-up, let's list the main steps that you can use to solve trig equations. First, reduce them to the form: $\sin x = a$, $\cos x = a$, or $\tan x = a$. Then, locate the solutions within the main interval. Finally, use the period of the function to find all the solutions.

Practice Problems

Problem 1: Solve $3\sec^2 x - 4 = 0$.

Problem 2: Solve $\sin^2 x = 3\cos^2 x$.

Problem 3: Solve $2\sin^2 x + 3\sin x + 1 = 0$ in the interval $[0, 2\pi]$.

Problem 4: Solve $\sin x = \cos x$. Hint: square both sides.

Problem 5: Solve $3\tan^3 x - 3\tan^2 x - \tan x + 1 = 0$. Hint: use factoring by grouping.

The Least You Need to Know

- Trigonometric functions are periodic functions; thus there are infinitely many solutions for trigonometric equations. Solutions that include all possible angles are called the general form of the solution.
- Factoring is one of the most common techniques for solving trigonometric equations.
- If a trigonometric equation in a quadratic form cannot be factored, use the quadratic formula.
- Checking your answers is a good practice to avoid mistakes. You can also use a graphing calculator to check your solutions.

Sum and Difference Formulas

In This Chapter

- Using identities with sums of angles
- Using identities with differences of angles
- Determining exact function values for more angles
- Solving trig equations using the sum and difference formulas

In this chapter, you study several trigonometric identities that are important in scientific applications. These are different from the fundamental identities that you studied in Chapter 14. These new identities help to find exact function values for a lot more angles by adding and subtracting the products of the function values for the special angles of 0°, 30°, 45°, 60°, and 90°.

Sum and Difference Formulas for Sine and Cosine

Have you ever found it unsettling that you can find the exact values of trigonometric functions for special angles but not, for example, for the angles of 15° or 75°? This is exactly what the identities that we are going to introduce do—they help to expand the pool of angles for which you can find the exact values of trigonometric functions.

These new identities are called the sum and difference formulas. These formulas allow you to break up the original angles into two special angles and find the function values of the original angle by using the function values of two special angles.

Let's consider several examples of how you can break some of the angles into the sums or differences of two special angles:

$$120° = 60° + 60° \text{ or } 120° = 90° + 30°$$

$75° = 30° + 45°$

$15° = 60° - 45°$

In this section, we consider the sum and difference formulas for the sine and cosine functions:

$$\sin(u + v) = \sin u \cos v + \cos u \sin v$$

$$\sin(u - v) = \sin u \cos v - \cos u \sin v$$

$$\cos(u + v) = \cos u \cos v - \sin u \sin v$$

$$\cos(u - v) = \cos u \cos v + \sin u \sin v$$

These formulas were derived by Euler, who, as we mentioned in Chapter 1, had a tremendous impact on trigonometry. They are derived via operations with complex numbers, which we will study further in Chapter 21 of this book.

Note that in the sine formulas, the + or − on the left is also a + or − on the right. But in the cosine formulas, the + on the left becomes a − on the right, and vice versa.

There are two main purposes of the sum and difference formulas: finding exact values of trigonometric functions and simplifying expressions to obtain other identities. The following problems illustrate how to use these formulas.

Sample Problem 1

Find the exact value of $\cos 75°$.

Step 1: Use the following fact:

$75° = 30° + 45°$

Step 2: Use the sum formula for the cosine:

$$\cos(u + v) = \cos u \cos v - \sin u \sin v$$

$$\cos 75° = \cos(30° + 45°) = \cos 30° \cos 45° - \sin 30° \sin 45°$$

Step 3: Plug in the values for the sines and cosines of special angles:

$$\cos 30° \cos 45° - \sin 30° \sin 45° = \frac{\sqrt{3}}{2} \cdot \frac{\sqrt{2}}{2} - \frac{1}{2} \cdot \frac{\sqrt{2}}{2} = \frac{\sqrt{6} - \sqrt{2}}{4}$$

You find that $\cos 75° = \dfrac{\sqrt{6} - \sqrt{2}}{4}$.

WORTH KNOWING

With sum and difference formulas, you can find the function values of any angle, but in this chapter you use only the most convenient combinations—ones with exact values that you can insert easily into the formulas.

In your next problem, you use the radian measures of the angles.

Sample Problem 2

Find the exact value of $\cos\dfrac{\pi}{12}$.

Step 1: Use the following fact:

$$\frac{\pi}{12} = \frac{\pi}{3} - \frac{\pi}{4}$$

Step 2: Use the difference formula for cosines:

$$\cos(u - v) = \cos u \cos v + \sin u \sin v$$

$$\cos\frac{\pi}{12} = \cos\left(\frac{\pi}{3} - \frac{\pi}{4}\right) = \cos\frac{\pi}{3}\cos\frac{\pi}{4} + \sin\frac{\pi}{3}\sin\frac{\pi}{4}$$

Step 3: Plug in the values of trig functions for special angles:

$$\cos\frac{\pi}{3}\cos\frac{\pi}{4} + \sin\frac{\pi}{3}\sin\frac{\pi}{4} = \frac{1}{2}\cdot\frac{\sqrt{2}}{2} + \frac{\sqrt{3}}{2}\cdot\frac{\sqrt{2}}{2} = \frac{\sqrt{2}+\sqrt{6}}{4}$$

Solution: You find that $\cos\dfrac{\pi}{12} = \dfrac{\sqrt{2}+\sqrt{6}}{4}$.

If it is difficult to figure out how to break angles into two special angles when they are given in radian measures, you can convert the original angle into degrees and determine how to break it up then. Afterward, you can convert the angles back to radians. For example, in your previous problem you might want to convert $\dfrac{\pi}{12}$ into a 15° angle (using conversion formulas that we discuss in Chapter 8 if you cannot figure its degree measure right away), then break it up as 15° = 60° − 45°, and finally convert everything back to radians.

Sum and Difference Formulas for the Tangent

Similar sum and difference formulas exist for the tangent function as well. These are the following:

$$\tan(u+v) = \frac{\tan u + \tan v}{1 - \tan u \tan v}$$

$$\tan(u-v) = \frac{\tan u - \tan v}{1 + \tan u \tan v}$$

DANGEROUS TURN

Only the main three trigonometric functions have user-friendly sum and difference formulas. The similar identities for the reciprocal functions are complicated. If you need the sum or difference of any of the reciprocal functions, use the corresponding fundamental identity and find the reciprocal of the answer after you are done.

The next two problems illustrate the use of the sum and difference formulas for the tangent.

Sample Problem 3

Find the exact value of $\dfrac{\tan 55° + \tan 80°}{1 - \tan 55° \tan 80°}$.

Step 1: Use the sum formula for the tangent:

$$\frac{\tan 55° + \tan 80°}{1 - \tan 55° \tan 80°} = \tan(55° + 80°) = \tan 135°$$

Step 2: An angle of 135° is not one of the special angles but you can find its reference angle, which is a special angle for which you know the values:

Reference angle = 180° − 135° = 45°

tan45° = 1

Step 3: The tangent value for 135° has the same numerical value as the angle of 45°. You need to decide on the sign. Because the angle 135° is a second-quadrant angle, its tangent has a negative value. Thus:

tan135° = −1

$$\frac{\tan 55° + \tan 80°}{1 - \tan 55° \tan 80°} = -1$$

Solution: The exact value of the given expression is equal to –1.

In your next problem, you use the radian measures of angles.

Sample Problem 4

Find the tangent of the angle $\dfrac{7\pi}{12}$.

Step 1: If it is difficult for you to figure out which two special angles in radians constitute the given angle, convert the original angle into degrees and find the sum of the special angles in degrees:

$$\frac{7\pi}{12} \cdot \frac{180}{\pi} = 105°$$

$$105° = 60° + 45°$$

WORTH KNOWING

Sometimes the sum and difference formulas are referred to as addition formulas.

Step 2: Rewrite the previous expression in radians:

$$\frac{7\pi}{12} = \frac{\pi}{3} + \frac{\pi}{4}$$

Step 3: Use the sum formula for the tangent and insert the tangent values for special angles:

$$\tan\frac{7\pi}{12} = \tan\left(\frac{\pi}{3} + \frac{\pi}{4}\right) = \frac{\tan\dfrac{\pi}{3} + \tan\dfrac{\pi}{4}}{1 - \tan\dfrac{\pi}{3}\tan\dfrac{\pi}{4}}$$

$$\frac{\sqrt{3}+1}{1-\sqrt{3}\cdot 1} = \frac{\sqrt{3}+1}{1-\sqrt{3}}$$

Solution: You find that $\tan\dfrac{7\pi}{12} = \dfrac{\sqrt{3}+1}{1-\sqrt{3}}$.

Applying Formulas to Various Trig Problems

In this section, we explore other types of problems that make use of the sum and difference formulas. These problems illustrate other uses of addition formulas that are different from the ones we previously discussed.

In your next problem, you use the formula to evaluate an expression.

Sample Problem 5

Find the value of $\cos(u - v)$ given that $\cos u = -\dfrac{15}{17}$, $\pi < u < \dfrac{3\pi}{2}$ and $\sin v = \dfrac{4}{5}$, $0 < v < \dfrac{\pi}{2}$.

Step 1: Before you can use the difference formula, you need to find the values of $\sin u$ and $\cos v$.

Find the value of $\sin u$ using the Pythagorean identity:

$$\sin^2 u + \cos^2 u = 1 \rightarrow \sin^2 u = 1 - \cos^2 u$$

$$\sin u = \pm\sqrt{1 - \cos^2 u} = \pm\sqrt{1 - \left(-\frac{15}{17}\right)^2} = \pm\sqrt{1 - \frac{225}{289}} = \pm\sqrt{\frac{64}{289}} = \pm\frac{8}{17}$$

Because, according to the problem's information, the $\angle u$ is a third-quadrant angle, you need to choose a negative value for the sine of $\angle u$.

$$\sin u = -\frac{8}{17}$$

Step 2: Similarly, you can find the value of $\cos v$ using the Pythagorean identity:

$$\sin^2 v + \cos^2 v = 1 \rightarrow \cos^2 v = 1 - \sin^2 v$$

$$\cos v = \pm\sqrt{1 - \sin^2 v} = \pm\sqrt{1 - \left(\frac{4}{5}\right)^2} = \pm\sqrt{1 - \frac{16}{25}} = \pm\sqrt{\frac{9}{25}} = \pm\frac{3}{5}$$

Because, according to the problem's information, $\angle v$ is a first-quadrant angle, you need to choose a positive value for the cosine of $\angle v$.

$$\cos v = \frac{3}{5}$$

Step 3: Use the difference formula to find the value for $\cos(u - v)$:

$$\cos(u - v) = \cos u \cos v + \sin u \sin v =$$

$$= \left(-\frac{15}{17}\right)\left(\frac{3}{5}\right) + \left(-\frac{8}{17}\right)\left(\frac{4}{5}\right) = -\frac{45}{85} - \frac{32}{85} = -\frac{77}{85}$$

Solution: You find that $\cos(u - v) = -\dfrac{77}{85}$.

You can also use the formulas to prove cofunction identities. The next problem illustrates how to do this.

Sample Problem 6

Prove the cofunction identity $\cos\left(\dfrac{\pi}{2} - x\right) = \sin x$.

Step 1: Use the difference formula:

$$\cos\left(\frac{\pi}{2} - x\right) = \cos\frac{\pi}{2}\cos x + \sin\frac{\pi}{2}\sin x$$

Step 2: Plug in the values of the functions for $\dfrac{\pi}{2}$:

$$\cos\left(\frac{\pi}{2} - x\right) = \cos\frac{\pi}{2}\cos x + \sin\frac{\pi}{2}\sin x = 0 \cdot \cos x + 1 \cdot \sin x = \sin x$$

Solution: You proved cofunction identity $\cos\left(\dfrac{\pi}{2} - x\right) = \sin x$.

Some of the trigonometric equations, such as the ones in the following problems, can be solved using sum and difference formulas.

Sample Problem 7

Solve the equation $\cos\left(x + \dfrac{\pi}{4}\right) + \cos\left(x - \dfrac{\pi}{4}\right) = 1$ in the interval $[0, 2\pi)$.

Step 1: Using sum and difference formulas, you can rewrite the left side of the equation as the following:

$$\cos\left(x + \frac{\pi}{4}\right) + \cos\left(x - \frac{\pi}{4}\right) =$$

$$= \cos x \cos\frac{\pi}{4} - \sin x \sin\frac{\pi}{4} + \cos x \cos\frac{\pi}{4} + \sin x \sin\frac{\pi}{4}$$

Step 2: After collecting like terms and plugging in the value for the cosine of $\frac{\pi}{4}$, you obtain:

$$2\cos x \cos\frac{\pi}{4} = 1 \text{ or } 2\cos x \cdot \frac{\sqrt{2}}{2} = 1$$

$$\sqrt{2}\cos x = 1 \text{ or } \cos x = \frac{1}{\sqrt{2}} = \frac{\sqrt{2}}{2}$$

The only angles that have this value for the cosine on the interval $[0, 2\pi)$ are $\frac{\pi}{4}$ and $\frac{7\pi}{4}$.

Solution: You find that $x = \frac{\pi}{4}$ and $x = \frac{7\pi}{4}$.

The last problem in this chapter makes use of an identity that you'll prove now. You want to verify that, for three angles of triangle A, B, and C, the following is true:

$$\sin C = \sin A\cos B + \cos A\sin B$$

To prove this, let's use the fact that the sum of a triangle's three angles is 180°, then $\angle C = 180° - (\angle A + \angle B)$. Therefore, you can write that:

$$\sin C = \sin\left[180° - (A + B)\right]$$

Use the difference formula for sines:

$$\sin C = \sin\left[180° - (A + B)\right] = \sin 180° \cos(A + B) - \cos 180° \sin(A + B)$$

Plug the values for the sine and cosine of 180°:

$$\sin C = \sin 180° \cos(A + B) - \cos 180° \sin(A + B) =$$
$$= 0 \cdot \cos(A + B) - (-1)\sin(A + B) = \sin(A + B)$$

Now use the sum formula for sines:

$$\sin C = \sin(A + B) = \sin A\cos B + \cos A\sin B$$

You proved the identity. Use it now for your last problem.

Sample Problem 8

The cosines of a triangle's two angles are $\cos A = \dfrac{1}{3}$ and $\cos B = \dfrac{2}{3}$. Find the sine of the third angle, $\angle C$.

Step 1: To find the sine of $\angle C$ you can use the identity that you just proved:

$$\sin C = \sin A \cos B + \cos A \sin B$$

But to find the numerical value for the sine of $\angle C$, you need to know the sines of $\angle A$ and $\angle B$.

Step 2: Knowing the cosine of $\angle A$, you can find its sine using the Pythagorean identity:

$$\sin^2 A = 1 - \cos^2 A$$

Extract the square root and substitute the value of the cosine:

$$\sin A = \pm\sqrt{1 - \cos^2 A} = \pm\sqrt{1 - \left(\frac{1}{3}\right)^2} = \pm\sqrt{1 - \frac{1}{9}} = \pm\sqrt{\frac{8}{9}} = \pm\frac{2\sqrt{2}}{3}$$

Discard the negative value, because only angles that are larger than 180° have negative sine values. These angles are impossible to have in any triangle. You find that $\sin A = \dfrac{2\sqrt{2}}{3}$.

Step 3: Similarly, you find the sine of $\angle B$. Knowing the cosine of $\angle B$, you can find its sine using the Pythagorean identity:

$$\sin^2 B = 1 - \cos^2 B$$

Extract the square root and substitute the value of the cosine:

$$\sin B = \pm\sqrt{1 - \cos^2 B} = \pm\sqrt{1 - \left(\frac{2}{3}\right)^2} = \pm\sqrt{1 - \frac{4}{9}} = \pm\sqrt{\frac{5}{9}} = \pm\frac{\sqrt{5}}{3}$$

You again discard the negative value for the same reasons. You find that $\sin B = \dfrac{\sqrt{5}}{3}$.

> **WORD OF ADVICE**
>
> When dealing with radicals, use the following rules:
>
> Quotient rule: $\sqrt{\dfrac{a}{b}} = \dfrac{\sqrt{a}}{\sqrt{b}}$
>
> Product rule: $\sqrt{a \cdot b} = \sqrt{a} \cdot \sqrt{b}$
>
> For example, using these rules, you get:
>
> $$\sqrt{\frac{8}{9}} = \frac{\sqrt{8}}{\sqrt{9}} = \frac{\sqrt{4 \cdot 2}}{3} = \frac{\sqrt{4} \cdot \sqrt{2}}{3} = \frac{2\sqrt{2}}{3}.$$

Step 4: Lastly, plug in the values from the problem and the function values that you found into the identity that you proved earlier to find the sine of $\angle C$:

$$\sin C = \sin A \cos B + \cos A \sin B$$

$$\sin C = \frac{2\sqrt{2}}{3} \cdot \frac{2}{3} + \frac{1}{3} \cdot \frac{\sqrt{5}}{3} = \frac{4\sqrt{2}}{9} + \frac{\sqrt{5}}{9} = \frac{4\sqrt{2} + \sqrt{5}}{9}$$

Solution: You find that $\sin C = \dfrac{4\sqrt{2} + \sqrt{5}}{9}$.

We recommend that you memorize these sum and difference formulas for future use, as they are utilized in solving many trigonometry problems. Actually, you need to memorize only three formulas for sums, and just remember how the signs change for the difference formulas.

Practice Problems

Problem 1: Find the cosine of 165°.

Problem 2: Verify the identity of $\sin\left(\dfrac{\pi}{2} + x\right) = \cos x$.

Problem 3: Find all solutions of the equation $\sin\left(x + \dfrac{\pi}{4}\right) + \sin\left(x - \dfrac{\pi}{4}\right) = -1$ in the interval $[0, 2\pi)$.

Problem 4: Find the tangent of 255°.

Problem 5: Find $\sin(v - u)$ if $\sin u = \dfrac{7}{25}, \dfrac{\pi}{2} < u < \pi$ and $\cos v = \dfrac{4}{5}, \dfrac{3\pi}{2} < v < 2\pi$.

The Least You Need to Know

- Sum and difference formulas for the sine are:

    $\sin(u + v) = \sin u \cos v + \cos u \sin v$

    $\sin(u - v) = \sin u \cos v - \cos u \sin v$

- Sum and difference formulas for the cosine are:

    $\cos(u + v) = \cos u \cos v - \sin u \sin v$

    $\cos(u - v) = \cos u \cos v + \sin u \sin v$

- Sum and difference formulas for the tangent are:

$$\tan(u + v) = \frac{\tan u + \tan v}{1 - \tan u \tan v}$$

$$\tan(u + v) = \frac{\tan u + \tan v}{1 + \tan u \tan v}$$

- Sum and difference formulas help to find function values for many more angles.

Double-Angle and Power-Reducing Formulas

In This Chapter

- Doubling and tripling angle values
- Proving new identities by using the previously learned ones
- Reducing the power of trigonometric functions
- Applying identities to various trig problems

In this chapter, we study two other categories of trigonometric identities. The first involves functions of multiple angles such as $\sin bu$ or $\cos bu$. The second involves squares of trigonometric functions such as $\sin^2 u$. You learn how to prove some of these identities and how to apply these new identities to solve various trigonometric problems.

Double-Angle Formulas

Let's first deal with double-angle formulas, which break an expression involving a single angle into one involving two angles, and prove at least one of them. The proof is short and easy, which is a good starting point for this chapter.

Prove that $\sin 2u = 2\sin u \cos u$.

To prove this identity, let $v = u$ in the sum formulas for sines. Then obtain the following:

$$\sin 2u = \sin(u + u) = \sin u \cos u + \cos u \sin u = 2\sin u \cos u$$

Note that the order of the factors does not change the product; thus $\sin u \cos u = \cos u \sin u$.

You proved your first double-angle formula. The other double-angle formulas can be proved in a similar way. Let's list all three of them.

The following are called double-angle formulas:

$$\sin 2u = 2\sin u\cos u$$

$$\cos 2u = \cos^2 u - \sin^2 u = 2\cos^2 u - 1 = 1 - 2\sin^2 u$$

$$\tan 2u = \frac{2\tan u}{1 - \tan^2 u}$$

Observe that the double-angle formula for the cosine has three versions. The second and the third versions can be easily derived from the first version using the Pythagorean identity. Let's illustrate how you can derive the third version from the first version.

You have $\cos 2u = \cos^2 u - \sin^2 u$.

Use the Pythagorean identity:

$$\cos 2u = \cos^2 u - \sin^2 u = (1 - \sin^2 u) - \sin^2 u = 1 - 2\sin^2 u$$

Having three versions for the cosine of a double-angle provides a big advantage because you can choose between the versions and use the one that is the most convenient for any particular case.

Now let's put formulas to work and illustrate how you can use them for different trigonometric problems.

The first sample problem illustrates how to evaluate functions involving double-angles.

Sample Problem 1

Find the sine, cosine, and tangent of the angle 2θ if $\cos\theta = \dfrac{5}{13}$ and $\dfrac{3\pi}{2} < \theta < 2\pi$.

Step 1: Find the sine using the Pythagorean identity:

$$\sin\theta = \pm\sqrt{1 - \cos^2\theta} = \pm\sqrt{1 - \left(\frac{5}{13}\right)^2} = \pm\sqrt{\frac{144}{169}} = \pm\frac{12}{13}$$

You use the negative value for the sine because the angle is in the fourth quadrant, where the sine is negative.

Step 2: Knowing the sine and cosine for angle θ, you can find the function's values for a double-angle. Find the sine value:

$$\sin 2\theta = 2\sin\theta\cos\theta = 2\left(-\frac{12}{13}\right)\left(\frac{5}{13}\right) = -\frac{120}{169}$$

Step 3: Find the cosine value using the double-angle formulas:

$$\cos 2\theta = 2\cos^2\theta - 1 = 2\left(\frac{5}{13}\right)^2 - 1 =$$

$$= 2\cdot\frac{25}{169} - 1 = \frac{50}{169} - \frac{169}{169} = -\frac{119}{169}$$

Step 4: Find the tangent value using the quotient identity:

$$\tan 2\theta = \frac{\sin 2\theta}{\cos 2\theta} = \frac{-\dfrac{120}{169}}{-\dfrac{119}{169}} = \frac{120\cdot 169}{119\cdot 169} = \frac{120}{119}$$

Solution: You find that $\sin 2\theta = -\dfrac{120}{169}$, $\cos 2\theta = -\dfrac{119}{169}$, and $\tan 2\theta = \dfrac{120}{119}$.

Your next problem shows how to locate *relative maximums* and *relative minimums* of trigonometric functions. Because the sine and cosine are periodic functions, they have infinitely many relative maximums and minimums, and we will show you how to find them specifically on the interval $[0, 2\pi]$. Other relative maximums and minimums can be found by adding the period of the function to the results you obtain.

DEFINITION

A function $f(x)$ has a **relative maximum** value at $x = a$ if $f(a)$ is greater than any value on a certain interval around it. It is called a relative maximum because other values of the function might in fact be even greater.

A function $f(x)$ has a **relative minimum** value at $x = b$ if $f(b)$ is less than any value on a certain interval around it. It is called a relative minimum because other values of the function might in fact be even less.

Sample Problem 2

Find the relative minimum and relative maximums of $y = 6\cos^2 x - 3$ in the interval $[0, 2\pi]$.

Step 1: Factor the common factor to get:

$$y = 6\cos^2 x - 3 = 3(2\cos^2 x - 1)$$

Step 2: Using a double-angle identity in the parentheses, you can rewrite the given function as:

$$y = 3(2\cos^2 x - 1) = 3\cos 2x$$

Step 3: Using techniques discussed in Chapter 11, you can recognize that the graph of this function has an amplitude of 3 and a period of π (it is shrunk along the x-axis). The key points in the interval $[0, 2\pi]$ are the following:

Maximum	Intercept	Minimum	Intercept	Maximum
$(0,3)$	$\left(\dfrac{\pi}{4}, 0\right)$	$\left(\dfrac{\pi}{2}, -3\right)$	$\left(\dfrac{3\pi}{4}, 0\right)$	$(\pi, 3)$

Solution: Relative maximums are at points $(0,3)$ and $(\pi, 3)$, and relative minimum is at $\left(\dfrac{\pi}{2}, -3\right)$.

Double-angle formulas also help to solve trigonometric equations, as the next problem illustrates.

Sample Problem 3

Solve $4\cos x + 2\sin 2x = 0$.

Step 1: To solve the equation, rewrite it so that it involves functions of x only (rather than $2x$). For this, use the double-angle formula:

$$4\cos x + 2\sin 2x = 0$$

$$4\cos x + 4\sin x\cos x = 0$$

DANGEROUS TURN

Note that $\sin 2u \neq 2\sin u$. Similar statements are true for other trigonometric functions—therefore, $\cos 2u \neq 2\cos u$ and $\tan 2u \neq 2\tan u$. Thus, if you know the value of $\sin u$, do not double the angle to find the value of $2\sin u$.

Step 2: Factor the common factor:

$$4\cos x + 4\sin x\cos x = 4\cos x(1 + \sin x) = 0$$

Step 3: Set each factor to zero and find solutions in the interval $[0, 2\pi]$:

$\cos x = 0$ \qquad\qquad $1 + \sin x = 0$ or $\sin x = -1$

$x = \dfrac{\pi}{2}$ and $\dfrac{3\pi}{2}$ \qquad $x = \dfrac{3\pi}{2}$

Step 4: Find the general solution:

$$x = \frac{\pi}{2} + 2n\pi \text{ and } x = \frac{3\pi}{2} + 2n\pi$$

Solution: You find that $x = \dfrac{\pi}{2} + 2n\pi$ and $x = \dfrac{3\pi}{2} + 2n\pi$.

The double-angle formulas are not restricted to angles $2u$ and u. Other double combinations, such as $4u$ and $2u$, are also valid. These are some examples:

$$\sin 4u = 2\sin 2u\cos 2u$$

$$\cos 6u = \cos^2 3u - \sin^2 3u$$

Sample Problem 4

Derive a formula for $\sin 4x$ in terms of functions of x.

Step 1: Rewrite angle $4x$ as $2(2x)$ and apply the double-angle formula the first time:

$$\sin 4x = \sin 2(2x) = 2\sin 2x\cos 2x$$

Step 2: Use the double-angle formula for the sine and cosine:

$$2\sin 2x\cos 2x = 2(2\sin x\cos x)(\cos^2 x - \sin^2 x) = 4\sin x\cos^3 x - 4\sin^3 x\cos x$$

Therefore, $\sin 4x = 4\sin x\cos^3 x - 4\sin^3 x\cos x$.

Solution: You find that $\sin 4x = 4\sin x\cos^3 x - 4\sin^3 x\cos x$.

Using other forms of double-angle formula for cosine, you get different—but mathematically equivalent—results.

Deriving a Triple-Angle Formula

Multiple-angle formulas can be derived not only for even multiples of angle x (such as $2x$, $4x$, or $6x$) but for odd multiples as well. Let's derive a triple-angle formula.

Consider that $3x = 2x + x$, then:

$$\sin3x = \sin(2x + x)$$

Use the sum formula for the sine:

$$\sin(2x + x) = \sin2x\cos x + \cos2x\sin x$$

Use double-angle formulas for the sine and cosine:

$$\sin2x\cos x + \cos2x\sin x = 2\sin x\cos x\cos x + (1 - 2\sin^2x)\sin x$$

Square the cosine in the first term, distribute the second term, and then use the Pythagorean identity for the cosine:

$$2\sin x\cos^2x + \sin x - 2\sin^3x$$
$$2\sin x(1 - \sin^2x) + \sin x - 2\sin^3x$$

Distribute the first term and collect like terms:

$$2\sin x - 2\sin^3x + \sin x - 2\sin^3x = 3\sin x - 4\sin^3x$$

You find that the triple-angle formula is the following:

$$\sin3x = 3\sin x - 4\sin^3x$$

Power-Reducing Formulas

The double-angle formulas can be used to obtain power-reducing formulas that help to solve many problems related to calculus. These formulas reduce an expression where the trigonometric function is squared into expressions where it is not.

The following are called power-reducing formulas:

$$\sin^2 u = \frac{1-\cos 2u}{2} \qquad \cos^2 u = \frac{1+\cos 2u}{2} \qquad \tan^2 u = \frac{1-\cos 2u}{1+\cos 2u}$$

Let's see how these formulas help you to simplify trigonometric expression.

Sample Problem 5

Rewrite $\cos^4 x$ as a sum involving only first powers of the sines of multiple angles.

Step 1: Let's rewrite the expression to first use squares instead of taking cosine to the power of 4:

$$\cos^4 x = (\cos^2 x)^2$$

Step 2: We can then use the power-reducing formula:

$$\cos^4 x = \left(\cos^2 x\right)^2 = \left(\frac{1+\cos 2x}{2}\right)^2$$

Step 3: Square a fraction and use the square-the-trinomial pattern in the numerator:

$$\left(\frac{1+\cos 2x}{2}\right)^2 = \frac{1+2\cos 2x+\cos^2 2x}{4}$$

Step 4: Use the power-reducing formula again for the third term of the numerator:

$$\frac{1+2\cos 2x+\cos^2 2x}{4} = \frac{1}{4}\left(1+2\cos 2x+\frac{1+\cos 4x}{2}\right) =$$

$$= \frac{1}{4}\left(1+2\cos 2x+\frac{1}{2}+\frac{\cos 4x}{2}\right)$$

Step 5: Multiply each term in parentheses by a fraction $\frac{1}{4}$:

$$\frac{1}{4}\left(1+2\cos 2x+\frac{1}{2}+\frac{\cos 4x}{2}\right) = \frac{1}{4}+\frac{\cos 2x}{2}+\frac{1}{8}+\frac{\cos 4x}{8}$$

Step 6: Finally, collect like terms:

$$\frac{1}{4}+\frac{\cos 2x}{2}+\frac{1}{8}+\frac{\cos 4x}{8} = \frac{3}{8}+\frac{\cos 2x}{2}+\frac{\cos 4x}{8} =$$

$$= \frac{1}{8}\left(3+4\cos 2x+\cos 4x\right)$$

Solution: You find that $\cos^4 x = \frac{1}{8}\left(3+4\cos 2x+\cos 4x\right)$.

The last problem of the chapter illustrates how you can use the power-reducing formulas to obtain the exact values of the sine squared for angles that are not special angles.

Sample Problem 6

Find the exact value of $\sin^2(157.5°)$.

Step 1: Use the power-reducing formula for the sine:

$$\sin^2(157.5°) = \frac{1 - \cos(2 \cdot 157.5°)}{2} = \frac{1 - \cos 315°}{2}$$

Step 2: The reference angle for angle of 315° is 45°. You need to use a positive cosine value because angle 315° is a fourth-quadrant angle, where the cosine is positive:

$$\cos 315° = \cos 45° = \frac{\sqrt{2}}{2}$$

Step 3: Plug in the value for the cosine into the power-reducing formula:

$$\sin^2(157.5°) = \frac{1 - \cos 315°}{2} = \frac{1 - \left(\dfrac{\sqrt{2}}{2}\right)}{2}$$

Simplify:

$$\sin^2(157.5°) = \frac{1 - \left(\dfrac{\sqrt{2}}{2}\right)}{2} = \frac{1 - \dfrac{\sqrt{2}}{2}}{2} = \frac{\dfrac{2 - \sqrt{2}}{2}}{2} = \frac{2 - \sqrt{2}}{4}$$

Thus, $\sin^2(157.5°) = \dfrac{2 - \sqrt{2}}{4}$.

Solution: You find that $\sin^2(157.5°) = \dfrac{2 - \sqrt{2}}{4}$.

You might wonder whether the trigonometric identities you learn in this chapter are used in real life. The answer is yes—and here is one example for you.

One important application is the wave pattern of a vibrating string. In addition to vibrating as a whole and producing a fundamental note, the string also vibrates in halves, thirds, and progressively smaller segments, producing overtones that are called harmonics. A trigonometric equation that describes this type of wave includes sums of sines and cosines of x, $2x$, $3x$, and greater multiples of x.

Although trigonometric calculations in practical applications are nowadays usually done by electronics rather than people (for example, computers and electronic musical instruments), knowledge of the fundamental mathematical methods is needed to design, program, and troubleshoot such systems.

Practice Problems

Problem 1: Find the exact value of the expression $\dfrac{2\sin\dfrac{\pi}{12}\cos\dfrac{\pi}{12}}{1-2\sin^2\dfrac{\pi}{12}}$.

Problem 2: Solve the equation $\sin x \cos x = 0.25$ on the interval $[0, 2\pi]$.

Problem 3: Find the exact value of the expression $\dfrac{2\tan 15°}{1-\tan^2 15°}$.

Problem 4: Find the exact values of $\sin 2u$, $\cos 2u$, and $\tan 2u$ if $\sin u = \dfrac{3}{5}$ and $0 < u < \dfrac{\pi}{2}$.

Problem 5: Use the power-reducing formula to rewrite the expression $\sin^4 x$ in terms of the first power of the cosine.

The Least You Need to Know

- The following are double-angle formulas:

  $\sin 2u = 2\sin u \cos u$

  $\cos 2u = \cos^2 u - \sin^2 u = 2\cos^2 u - 1 = 1 - 2\sin^2 u$

  $\tan 2u = \dfrac{2\tan u}{1-\tan^2 u}$

- Doubling the value of $\sin u$ does not provide the value for $\sin 2u$. Similar statements for other trig functions are true as well.

- The triple-angle formula is as follows:

  $\sin 3x = 3\sin x - 4\sin^3 x$

- Power-reducing formulas are:

  $\sin^2 u = \dfrac{1-\cos 2u}{2}$

  $\cos^2 u = \dfrac{1+\cos 2u}{2}$

  $\tan^2 u = \dfrac{1-\cos 2u}{1+\cos 2u}$

Half-Angle and Product-to-Sum Formulas

In This Chapter

- Taking half an angle and finding functions' values
- Observing the signs of functions of half-angles
- Expressing a product of trigonometric functions as a sum
- Writing a sum of trigonometric functions as a product

In this chapter, you continue your study of various trigonometric formulas. First we introduce formulas that involve functions of half-angles such as $\sin\dfrac{u}{2}$, and then you learn formulas that involve products of trigonometric functions such as $\sin u \cos u$.

Half-Angle Formulas

Similar to the trigonometric identities that we studied in the previous chapters, half-angle formulas are useful in transforming and simplifying trigonometric expressions. You can derive the half-angle formulas from double-angle formulas for cosine by replacing $\angle u$ with $\dfrac{u}{2}$.

$$\cos 2u = 1 - 2\sin^2 u$$

$$\cos 2\left(\frac{u}{2}\right) = 1 - 2\sin^2\frac{u}{2}$$

$$\cos u = 1 - 2\sin^2\frac{u}{2}$$

$$2\sin^2\frac{u}{2} = 1 - \cos u$$

$$\sin\frac{u}{2} = \pm\sqrt{\frac{1 - \cos u}{2}}$$

To derive the new formula for the cosine, use another version of a double-angle formula for the cosine:

$$\cos 2u = 2\cos^2 u - 1$$

$$\cos 2\left(\frac{u}{2}\right) = 2\cos^2\frac{u}{2} - 1$$

$$\cos u = 2\cos^2\frac{u}{2} - 1$$

$$2\cos^2\frac{u}{2} = 1 + \cos u$$

$$\cos\frac{u}{2} = \pm\sqrt{\frac{1 + \cos u}{2}}$$

The formulas for the tangent can be verified using the quotient identity. You get the following formulas:

$$\tan\frac{u}{2} = \pm\sqrt{\frac{1 - \cos u}{1 + \cos u}} = \frac{\sin u}{1 + \cos u} = \frac{1 - \cos u}{\sin u}$$

Because these formulas involve half-angles, they are called half-angle formulas. Here is a full list of all of them together:

$$\sin\frac{u}{2} = \pm\sqrt{\frac{1 - \cos u}{2}}$$

$$\cos\frac{u}{2} = \pm\sqrt{\frac{1 + \cos u}{2}}$$

$$\tan\frac{u}{2} = \pm\sqrt{\frac{1 - \cos u}{1 + \cos u}} = \frac{\sin u}{1 + \cos u} = \frac{1 - \cos u}{\sin u}$$

Notice that the half-angle formula for the tangent function can be expressed in three different ways, all equivalent to one another. The second and the third versions don't have the ± sign. Let's prove the second version to understand how it works.

As we said, use the quotient identity to derive the formula and plug in the half-angle identities for the sine and cosine.

$$\tan\frac{u}{2} = \frac{\sin\dfrac{u}{2}}{\cos\dfrac{u}{2}} = \frac{\sqrt{\dfrac{1 - \cos u}{2}}}{\sqrt{\dfrac{1 + \cos u}{2}}}$$

Notice that the ± signs were dropped because they are going to be squared out later anyway. Put the numerator and denominator under one radical and simplify the complex fraction:

$$\frac{\sqrt{\dfrac{1-\cos u}{2}}}{\sqrt{\dfrac{1+\cos u}{2}}} = \sqrt{\frac{\dfrac{1-\cos u}{2}}{\dfrac{1+\cos u}{2}}} = \sqrt{\frac{1-\cos u}{1+\cos u}}$$

To get rid of the radicals, multiply the numerator and the denominator by the conjugate of the denominator:

$$= \sqrt{\frac{(1-\cos u)(1-\cos u)}{(1+\cos u)(1-\cos u)}} = \sqrt{\frac{(1-\cos u)^2}{1-\cos^2 u}}$$

Note that you use the difference of two squares pattern in the denominator. Replace the denominator using the Pythagorean identity and then separate the numerator and denominator by putting them under separate radical signs.

$$\sqrt{\frac{(1-\cos u)^2}{1-\cos^2 u}} = \frac{\sqrt{(1-\cos u)^2}}{\sqrt{\sin^2 u}} = \frac{1-\cos u}{\sin u}$$

The third version of the half-angle formula for the tangent can be derived similarly if you multiply both the numerator and the denominator by the conjugate of the numerator instead of the denominator.

Having three versions of the same formula helps because you can choose the most convenient version depending on the situation and use the formula that makes solving the problem easier.

DANGEROUS TURN

Don't forget to identify the quadrants in which $\dfrac{u}{2}$ lies and use the signs accordingly for half-angle formulas that involve the sine and cosine.

The half-angle formulas for the sine and cosine are unique in terms of having ± signs attached to them. When you apply these formulas, you need to consider which quadrant angle u and $\dfrac{u}{2}$ are in and use the appropriate signs. For example, angle of 240° is a third-quadrant angle with a negative sine, but its half-angle, 120°, is a second-quadrant angle with a positive sine.

The next problems make use of half-angle formulas.

Sample Problem 1

Find the exact value of the cosine of 105°.

Step 1: Note that 105° is half of 210°. Use the half-angle formula:

$$\cos 105° = \pm\sqrt{\frac{1+\cos 210°}{2}}$$

Step 2: The 105° angle lies in quadrant II; therefore, use the negative sign because the cosine is negative in the second quadrant:

$$\cos 105° = -\sqrt{\frac{1+\cos 210°}{2}}$$

Step 3: Also, observe that a 30° angle is a reference angle for 210°; because 210° is a third-quadrant angle, use a negative sign for the cosine:

$$\cos 105° = -\sqrt{\frac{1+\cos 210°}{2}} = -\sqrt{\frac{1+(-\cos 30°)}{2}} = -\sqrt{\frac{1-\frac{\sqrt{3}}{2}}{2}} = -\sqrt{\frac{2-\sqrt{3}}{4}} = -\frac{\sqrt{2-\sqrt{3}}}{2}$$

Solution: $\cos 105° = -\dfrac{\sqrt{2-\sqrt{3}}}{2}$

Now use the half-angle formulas to solve trigonometric equations.

Sample Problem 2

Solve $2-\sin^2 x = 2\cos^2 \dfrac{x}{2}$.

Step 1: Use the half-angle formula for the cosine:

$$2-\sin^2 x = 2\cos^2 \frac{x}{2}$$

$$2-\sin^2 x = 2\left(\frac{1+\cos x}{2}\right)$$

Step 2: Simplify by distributing on the right:

$$2 - \sin^2 x = 1 + \cos x$$

Step 3: Use the Pythagorean identity, get all terms on the left side, and collect like terms:

$$2 - (1 - \cos^2 x) = 1 + \cos x$$

$$2 - 1 + \cos^2 x - 1 - \cos x = 0$$

$$\cos^2 x - \cos x = 0$$

Step 4: Factor the expression and equal each factor to zero to find the solutions for the interval $[0, 2\pi)$:

$$\cos x(\cos x - 1) = 0$$

$$\cos x = 0 \qquad\qquad\qquad \cos x - 1 = 0 \text{ or } \cos x = 1$$

$$x = \frac{\pi}{2} \text{ and } x = \frac{3\pi}{2} \qquad x = 0$$

Step 5: Therefore, the general solution is $x = 2n\pi$, $x = \frac{\pi}{2} + 2n\pi$, and $x = \frac{3\pi}{2} + 2n\pi$.

Step 6: The second and third solutions can be further combined and simplified to become $x = \frac{\pi}{2} + n\pi$. This is due to the fact that $\frac{\pi}{2} + n\pi$ can become equal to $\frac{3\pi}{2}$ if $n = 1$.

Solution: $x = 2n\pi$ and $x = \frac{\pi}{2} + n\pi$

Product–to–Sum Formulas

Another type of formula is called product-to-sum formulas. These can be easily verified using the sum and difference formulas discussed in the preceding chapter.

The following are the product-to-sum formulas:

$$\sin u \sin v = \frac{1}{2}\left[\cos(u - v) - \cos(u + v)\right]$$

$$\cos u \cos v = \frac{1}{2}\left[\cos(u - v) + \cos(u + v)\right]$$

$$\sin u \cos v = \frac{1}{2}\left[\sin(u + v) + \sin(u - v)\right]$$

$$\cos u \sin v = \frac{1}{2}\left[\sin(u + v) - \sin(u - v)\right]$$

Product-to-sum formulas allow you to rewrite products of trigonometric functions as a sum or difference of trigonometric functions, as the next sample problem illustrates.

Sample Problem 3

Rewrite $\cos 6x \sin 4x$ as a sum or difference.

Step 1: Use the product-to-sum formula for the product of the cosine and sine:

$$\cos u \sin v = \frac{1}{2}\left[\sin(u + v) - \sin(u - v)\right]$$

$$\cos 6x \sin 4x = \frac{1}{2}\left[\sin(6x + 4x) - \sin(6x - 4x)\right]$$

Step 2: Add the expressions inside the parentheses and multiply by a factor of $\frac{1}{2}$:

$$\frac{1}{2}\left[\sin(6x+4x)-\sin(6x-4x)\right]=\frac{1}{2}\sin 10x-\frac{1}{2}\sin 2x$$

Solution: You find that $\cos 6x\sin 4x=\frac{1}{2}\sin 10x-\frac{1}{2}\sin 2x$.

Sum-to-Product Formulas

Sometimes, it is useful to reverse the procedure and write a sum of trigonometric functions as a product. Formulas that are called sum-to-product formulas can help to accomplish that task.

The following are the sum-to-product formulas:

$$\sin u+\sin v=2\sin\left(\frac{u+v}{2}\right)\cos\left(\frac{u-v}{2}\right)$$

$$\sin u-\sin v=2\cos\left(\frac{u+v}{2}\right)\sin\left(\frac{u-v}{2}\right)$$

$$\cos u+\cos v=2\cos\left(\frac{u+v}{2}\right)\cos\left(\frac{u-v}{2}\right)$$

$$\cos u-\cos v=-2\sin\left(\frac{u+v}{2}\right)\sin\left(\frac{u-v}{2}\right)$$

Sample Problem 4

Find the exact value of sin195° + sin105°.

Step 1: Use the appropriate sum-to-product formula to obtain the following:

$$\sin 195°+\sin 105°=2\sin\left(\frac{195°+105°}{2}\right)\cos\left(\frac{195°-105°}{2}\right)$$

Step 2: Perform operations inside each pair of parentheses:

$$2\sin\left(\frac{195°+105°}{2}\right)\cos\left(\frac{195°-105°}{2}\right)=2\sin 150°\cos 45°$$

Step 3: Plug in the values of trigonometric functions. A reference angle for 150° is the angle of 30°. Use a positive value because 150° is a second-quadrant angle, where the sine is positive:

$$2\sin 150°\cos 45°=2\sin 30°\cos 45°=2\left(\frac{1}{2}\right)\left(\frac{\sqrt{2}}{2}\right)=\frac{\sqrt{2}}{2}$$

Therefore, $\sin 195° + \sin 105° = \dfrac{\sqrt{2}}{2}$.

Solution: You find that $\sin 195° + \sin 105° = \dfrac{\sqrt{2}}{2}$.

In the next problem, sum-to-product formulas help to solve a trigonometric equation.

Sample Problem 5

Solve $\sin 4x + \sin 2x = 0$ on the interval $[0, 2\pi)$, which means from 0 to 2π but not including 2π itself.

Step 1: Use sum-to-product formula:

$$\sin 4x + \sin 2x = 0$$

$$\sin 4x + \sin 2x = 2\sin\left(\frac{4x + 2x}{2}\right)\cos\left(\frac{4x - 2x}{2}\right) = 0$$

Step 2: Simplify and set each factor equal to zero to find solutions:

$$2\sin\left(\frac{6x}{2}\right)\cos\left(\frac{2x}{2}\right) = 2\sin 3x \cos x = 0$$

$$\sin 3x = 0 \qquad\qquad \cos x = 0$$

The solutions for the first equation $\sin 3x = 0$ are:

$$3x = 0 \text{ and } 3x = \pi, \text{ or } x = 0 + \frac{2n\pi}{3} \text{ and } x = \frac{\pi}{3} + \frac{2n\pi}{3}$$

Then, in the interval $[0, 2\pi)$, the solutions are:

$$x = 0, \ x = \frac{\pi}{3}, \ x = \frac{2\pi}{3}, \ x = \pi, \ x = \frac{4\pi}{3}, \text{ and } x = \frac{5\pi}{3}$$

For the second equation, the solutions in the interval $[0, 2\pi)$ are:

$$\cos x = 0, \text{ or } x = \frac{\pi}{2} \text{ and } x = \frac{3\pi}{2}$$

Solution: You find that $x = 0$, $x = \dfrac{\pi}{3}$, $x = \dfrac{2\pi}{3}$, $x = \pi$, $x = \dfrac{4\pi}{3}$, $x = \dfrac{5\pi}{3}$, $x = \dfrac{\pi}{2}$, and $x = \dfrac{3\pi}{2}$.

You can check solutions using a graphing calculator and sketching a graph.

The last problem illustrates how you can verify the identity using sum-to-product formulas.

Sample Problem 6

Verify the identity $\dfrac{\cos 4x + \cos 2x}{\sin 4x + \sin 2x} = \cot 3x$.

Step 1: Use sum-to-product formulas for both the numerator and denominator:

$$\frac{\cos 4x + \cos 2x}{\sin 4x + \sin 2x} = \frac{2\cos\left(\dfrac{4x + 2x}{2}\right)\cos\left(\dfrac{4x - 2x}{2}\right)}{2\sin\left(\dfrac{4x + 2x}{2}\right)\cos\left(\dfrac{4x - 2x}{2}\right)}$$

Step 2: Simplify inside each parentheses to obtain:

$$\frac{2\cos 3x \cos x}{2\sin 3x \cos x}$$

Step 3: Reduce and use the fundamental identity:

$$\frac{2\cos 3x \cos x}{2\sin 3x \cos x} = \frac{\cos 3x}{\sin 3x} = \cot 3x$$

Solution: You verified that $\dfrac{\cos 4x + \cos 2x}{\sin 4x + \sin 2x} = \cot 3x$.

How to Avoid Frustration While Working with Formulas

We want to conclude this chapter by addressing a concern that many students have about a great number of formulas to understand, remember, and use. It seems that there are just too many formulas to handle. There are shortcuts that enable you to remember formulas better. Now, let's count and tally the functions:

Sum and difference formulas	6
Double-angle formulas	3
Power-reducing formulas	3
Half-angle formulas	3
Product-to-sum formulas	4
Sum-to-product formulas	4
Total amount of formulas	23

There are six types of formulas, and you should memorize not only the formulas but also their names—because each name has a hint of what that type of formula is about. For example, the sum-to-product name clearly indicates that you start with the sum and end with the product of trigonometric functions.

Also, exclude formulas that involve the tangent. Thus, we are talking about only the formulas that involve the sine and cosine.

The following are memory aids that might help you to remember all these formulas:

- All formulas of three types that contain the word *sum* contain two angles, $\angle u$ and $\angle v$.

- All formulas from the types that don't contain this word have only one angle, $\angle u$, involved.

- In all formulas that contain fractions, the denominator is always 2.

- The formulas of all types contain only two varieties of trigonometric functions—the sine and cosine.

- Only two types of formulas—sum and difference and double-angle formulas—don't have fractions involved.

- The other four types of formulas have some distinct features that distinguish them from all other formulas. Let's identify these unique features:

  Half-angle formulas have a ± sign and the square root in them.

  Product-to-sum formulas have the fraction $\dfrac{1}{2}$ in front of brackets that include two sets of parentheses.

  Sum-to-product formulas contain a factor 2 on the left side, which is always a product of the sine and cosine of half the sum or difference of two angles, $\angle u$ and $\angle v$.

  Power-reducing formulas have a squared trigonometric function on the left.

Practice Problems

Problem 1: Find the value of $\sin\dfrac{x}{2}$ if $\cos x = \dfrac{12}{13}$ and $0 < x < \dfrac{\pi}{2}$.

Problem 2: Determine the exact value of the tangent of $\dfrac{\pi}{8}$.

Problem 3: Use the product-to-sum formulas to write $6\sin\dfrac{\pi}{4}\cos\dfrac{\pi}{4}$ as a sum.

Problem 4: Verify the identity $\dfrac{\cos 4x - \cos 2x}{2\sin 3x} = -\sin x$.

Problem 5: Solve the equation $\sin 6x - \sin 4x = 0$ in the interval $[0, 2\pi)$.

The Least You Need to Know

- The half-angle formulas are as follows:

$$\sin\frac{u}{2} = \pm\sqrt{\frac{1-\cos u}{2}}$$

$$\sin\frac{u}{2} = \pm\sqrt{\frac{1+\cos u}{2}}$$

$$\tan\frac{u}{2} = \pm\sqrt{\frac{1-\cos u}{1+\cos u}} = \frac{\sin u}{1+\cos u} = \frac{1-\cos u}{\sin u}$$

- When using half-angle formulas, observe the signs of the functions by deciding in which quadrant a specific half-angle lies.

- The product-to-sum formulas are as follows:

$$\sin u \sin v = \frac{1}{2}\left[\cos(u-v) - \cos(u+v)\right]$$

$$\cos u \cos v = \frac{1}{2}\left[\cos(u-v) + \cos(u+v)\right]$$

$$\sin u \cos v = \frac{1}{2}\left[\sin(u+v) + \sin(u-v)\right]$$

$$\cos u \sin v = \frac{1}{2}\left[\sin(u+v) - \sin(u-v)\right]$$

- The sum-to-product formulas are as follows:

$$\sin u + \sin v = 2\sin\left(\frac{u+v}{2}\right)\cos\left(\frac{u-v}{2}\right)$$

$$\sin u - \sin v = 2\cos\left(\frac{u+v}{2}\right)\sin\left(\frac{u-v}{2}\right)$$

$$\cos u + \cos v = 2\cos\left(\frac{u+v}{2}\right)\cos\left(\frac{u-v}{2}\right)$$

$$\cos u - \cos v = -2\sin\left(\frac{u+v}{2}\right)\sin\left(\frac{u-v}{2}\right)$$

Polar Coordinates and Complex Numbers

The first two chapters in this part are devoted to complex numbers. You learn how to present complex numbers geometrically and how to convert a complex number from its rectangular form to its polar form and vice versa.

We discuss how to multiply complex numbers in polar form and introduce you to new formulas that will help you find powers and roots of complex numbers.

The last chapter is about graphing calculators and how they can be used to perform tasks related to trigonometry. You learn how to evaluate trigonometric functions, solve trigonometric equations, find inverse trigonometric functions, and use your calculator's graphing utility to sketch graphs of trig functions.

Polar Coordinates

Chapter 20

In This Chapter

- Understanding the differences between two coordinate systems
- Finding more than one pair of polar coordinates for (*x,y*)
- Converting polar coordinates to rectangular ones and vice versa
- Writing polar equations in rectangular form and vice versa
- Sketching polar graphs

Chapters 3 and 9 discuss that the position of a point P in the plane can be described by an ordered pair of numbers called coordinates, which are the signed distances from the point of origin of the Cartesian coordinate system to point P. This chapter introduces another way of describing a point's position in the plane using polar coordinates.

Introducing Polar Coordinates

To form the *polar coordinate system* on a plane, fix a point O, called the pole or origin, and construct from O an initial ray called the polar axis, as shown in the following illustration. Each point P in the plane is assigned polar coordinates (r, θ) according to the following:

r = distance from O to P

θ = angle, counterclockwise from polar axis to segment OP

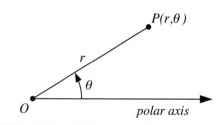

 DEFINITION

The **polar coordinate system** is a system where a point *P* in the plane is described by an ordered pair (r, *θ*), where *r* is the distance from the pole *O* to *P*, and *θ* is the measure of an angle formed by ray *OP* and the polar axis.

The numbers r and θ are called polar coordinates. The angle θ can be measured in both degrees and radians.

Plotting Points in the Polar Coordinate System

In the Cartesian coordinate system, a point is given by coordinates (x,y), and you plot the point by starting at the origin and then moving x units left and right horizontally followed by y units up and down vertically.

Let's now show how you can identify the same point differently using polar coordinates. Instead of moving vertically and horizontally from the origin to get to the point, you instead go straight out from the origin until you reach the point P and then determine the angle this line makes with the positive x-axis. You can then use the distance r of the point from the origin and the angle θ that you need to rotate from the positive x-axis as the coordinates of the point. This is shown in the following figure.

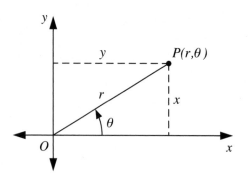

The previous discussion might lead you to think that r must be a positive number. However, r can be negative. The following figure illustrates two points $\left(2, \dfrac{\pi}{3}\right)$ and $\left(-2, \dfrac{\pi}{3}\right)$. To find a point where r is negative, simply find the ray that forms the angle θ with the polar axis, and then go r units in the opposite direction from the ray.

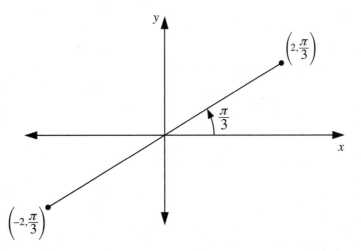

You can see that if r is positive, the point will be in the same quadrant as angle θ. But if r is negative, the point will end up in the quadrant exactly opposite angle θ.

Notice as well that the coordinates $\left(-2, \dfrac{\pi}{3}\right)$ describe the same point as the coordinates $\left(2, \dfrac{4\pi}{3}\right)$ do. These new coordinates dictate that you rotate an angle $\dfrac{4\pi}{3}$ from the positive x-axis and then move out a distance 2.

This discussion leads to an important difference between the rectangular and polar coordinate systems. In the rectangular coordinate system, each point has a unique representation with its coordinates (x,y). For example, point $(3,-5)$ cannot be described by any other ordered pair. This is not true for polar coordinates. In polar coordinates, there is an infinite number of coordinates for a given point.

Let's explain this statement by solving a sample problem.

Sample Problem 1

Plot the point $\left(5, \dfrac{\pi}{3}\right)$ and find three additional polar representations of this point, using the range $-2\pi < \theta < 2\pi$.

Step 1: If you rotate in a clockwise direction to get to the point, the angle is $-\dfrac{5\pi}{3}$, so the coordinates of the point are $\left(5, -\dfrac{5\pi}{3}\right)$ as the following figure illustrates.

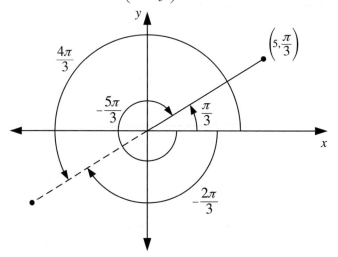

Step 2: As you rotate, if you end up in the opposite quadrant from the point, you can use a negative r to get back to the point. There is both a counterclockwise and a clockwise rotation to get to the angle.

If you move counterclockwise, the angle is $\dfrac{4\pi}{3}$, and the coordinates are $\left(-5, \dfrac{4\pi}{3}\right)$. If you move clockwise, then the angle is $-\dfrac{2\pi}{3}$, and the coordinates are $\left(-5, -\dfrac{2\pi}{3}\right)$.

Solution: Three other representations of the point are $\left(5, -\dfrac{5\pi}{3}\right)$, $\left(-5, \dfrac{4\pi}{3}\right)$, and $\left(-5, -\dfrac{2\pi}{3}\right)$.

> ✏️ **WORD OF ADVICE**
>
> In general, a point in the polar system can be represented as $(r, \theta \pm 2n\pi)$ or $(-r, \theta \pm (2n + 1)\pi)$, where n is any integer.

The four points that you find in the sample problem only represent the coordinates of the point without rotating around the system more than once. If you allow the angle to fulfill as many complete rotations about the axis system as you want, then there are an infinite number of coordinates for the same point.

Coordinate Conversions

To establish relationships between rectangular and polar coordinates, let's refer to the following illustration. Using the Pythagorean theorem, you can write:

$$r^2 = x^2 + y^2$$

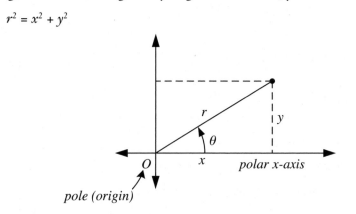

Using the trigonometric functions, you can write:

$$\cos\theta = \frac{x}{r} \text{ or } x = r\cos\theta \text{ and } \sin\theta = \frac{y}{r} \text{ or } y = r\sin\theta$$

WORTH KNOWING

The polar coordinates (r, θ) are related to the rectangular coordinates (x,y) as follows: $x = r\cos\theta$, $y = r\sin\theta$, $\tan\theta = \frac{y}{x}$, and $r^2 = x^2 + y^2$.

These relationships enable you to convert coordinates or equations from one system of coordinates to the other.

Polar-to-Rectangular Conversions

The conversion from polar coordinates back to rectangular coordinates is simple. All you have to do is use the expressions for x and y and plug in the values for r and the trigonometric functions of the given angle. The next two sample problems illustrate the procedure.

Sample Problem 2

Convert $\left(\sqrt{3}, \dfrac{\pi}{6}\right)$ to rectangular coordinates.

Step 1: Use the conversion formulas previously discussed:

$$x = r\cos\theta$$

$$y = r\sin\theta$$

Step 2: Plug in the values for r and the sine and cosine of angle θ:

$$x = r\cos\theta = \sqrt{3}\cdot\cos\frac{\pi}{6} = \sqrt{3}\cdot\frac{\sqrt{3}}{2} = \frac{3}{2}$$

$$y = r\sin\theta = \sqrt{3}\cdot\sin\frac{\pi}{6} = \sqrt{3}\cdot\frac{1}{2} = \frac{\sqrt{3}}{2}$$

Solution: The rectangular coordinates are $(x, y) = \left(\dfrac{3}{2}, \dfrac{\sqrt{3}}{2}\right)$.

Sample Problem 3

Convert $\left(4, -\dfrac{\pi}{3}\right)$ to the rectangular coordinates.

Step 1: Use the conversion formulas previously discussed:

$$x = r\cos\theta$$

$$y = r\sin\theta$$

Step 2: Because angle θ lies in the fourth quadrant, its sine is negative and its cosine is positive, so you use the function's values for angle $\dfrac{\pi}{3}$ with the appropriate signs.

Step 3: Plug in the values for r and the sine and cosine of angle θ:

$$x = r\cos\theta = 4\cdot\cos\left(-\frac{\pi}{3}\right) = 4\cdot\left(\frac{1}{2}\right) = 2$$

$$y = r\sin\theta = 4\cdot\sin\left(-\frac{\pi}{3}\right) = 4\cdot\left(-\frac{\sqrt{3}}{2}\right) = -2\sqrt{3}$$

Solution: The rectangular coordinates are $(x, y) = \left(2, -2\sqrt{3}\right)$.

Rectangular-to-Polar Conversions

To convert from rectangular coordinates to polar coordinates, you need to use the formulas for the tangent of an angle and for the radius. You can see how it works by doing sample problems.

Sample Problem 4

Convert $(x,y) = (-1,1)$ to polar coordinates.

Step 1: Use the conversion formulas previously discussed:

$$\tan\theta = \frac{y}{x}$$
$$r^2 = x^2 + y^2$$

Step 2: Find the tangent by plugging in the x and y values:

$$\tan\theta = \frac{y}{x} = \frac{1}{-1} = -1$$

Special angle $\frac{\pi}{4}$ has 1 as the value for the tangent. Because the obtained value of the tangent is with a negative sign, the angle must be in the second quadrant where the tangent is negative. Therefore, angle $\theta = \frac{3\pi}{4}$.

Step 3: Because angle θ lies in the same quadrant as (x, y), use positive r:

$$r = \sqrt{x^2 + y^2} = \sqrt{(-1)^2 + (1)^2} = \sqrt{2}$$

Solution: One set of polar coordinates is $(r,\theta) = \left(\sqrt{2}, \frac{3\pi}{4}\right)$.

Do one more problem on conversion from rectangular coordinates to polar coordinates.

WORD OF ADVICE

In polar coordinates, the pole is represented by $(0, \theta)$, where θ is any angle.

Sample Problem 5

Convert $(x, y) = \left(-\sqrt{3}, -\sqrt{3}\right)$ to polar coordinates. Find two sets of coordinates.

Step 1: Use the conversion formulas previously discussed:

$$\tan\theta = \frac{y}{x}$$
$$r^2 = x^2 + y^2$$

Step 2: Find the tangent by plugging in the x and y values:

$$\tan\theta = \frac{y}{x} = \frac{-\sqrt{3}}{-\sqrt{3}} = 1$$

Therefore, angle $\theta = \dfrac{\pi}{4}$.

Step 3: Because angle θ does not lie in the same quadrant as (x,y), use negative r:

$$r = -\sqrt{x^2 + y^2} = -\sqrt{\left(-\sqrt{3}\right)^2 + \left(-\sqrt{3}\right)^2} = -\sqrt{3+3} = -\sqrt{6}$$

Thus, one set of polar coordinates is $(r,\theta) = \left(-\sqrt{6}, \dfrac{\pi}{4}\right)$.

Step 4: To find another set of polar coordinates, use the formula $\left(-r, \theta \pm (2n+1)\pi\right)$ discussed previously and assume that $n = 0$. Then,

$$\left(-r, \theta \pm (2n+1)\pi\right) = \left(-\left(-\sqrt{6}\right), \dfrac{\pi}{4} \pm (2 \cdot 0 + 1)\pi\right) = \left(\sqrt{6}, \dfrac{\pi}{4} + \pi\right) = \left(\sqrt{6}, \dfrac{5\pi}{4}\right).$$

Solution: One set of polar coordinates is $(r,\theta) = \left(-\sqrt{6}, \dfrac{\pi}{4}\right)$.
Another possible set is $(r,\theta) = \left(\sqrt{6}, \dfrac{5\pi}{4}\right)$.

Converting Polar Equations to Rectangular Form

From the previous four sample problems, you can see that conversion from the polar coordinates to the rectangular coordinates was simple and straightforward, whereas conversion from the rectangular coordinates to the polar coordinates was more involved and required some additional considerations about the angle's quadrant. Also, as you probably noticed, you never expect two sets of answers for the first type of conversion, whereas the second conversion enables you to get as many sets of polar coordinates as you want.

For equation conversion, the opposite is true. To convert a rectangular equation to polar form, simply replace x by $r\cos\theta$ and y by $r\sin\theta$. Let's demonstrate the process with a sample problem.

Sample Problem 6

Convert the rectangular equation $y = x^2$ into polar form.

Step 1: Use the formulas:

$$x = r\cos\theta$$

$$y = r\sin\theta$$

Step 2: Replace x and y with the expressions from the formulas:

$$y = x^2 \rightarrow r\sin\theta = (r\cos\theta)^2$$

You obtain the equation in polar form.

Solution: The polar form of the equation $y = x^2$ is $r\sin\theta = (r\cos\theta)^2$.

On the other hand, the opposite equation conversion from polar form to the rectangular form can be challenging.

Sample Problem 7

Convert equation $r = \sec\theta$ to rectangular form.

Step 1: Revise the polar equation by using the reciprocal identity to get:

$$r = \sec\theta \rightarrow r = \frac{1}{\cos\theta} \text{ or } r\cos\theta = 1$$

Step 2: Compare the conversion formula $x = r\cos\theta$ with the previous expression:

$$r\cos\theta = 1 \qquad\qquad x = r\cos\theta$$

Because the left side of the first is equal to the right side of the other, you can conclude that:

$$x = 1 \rightarrow \text{This is the rectangular form of the original polar equation } r = \sec\theta.$$

Solution: The polar equation $r = \sec\theta$ is $x = 1$ in rectangular form.

WORTH KNOWING

The actual term *polar coordinates* has been attributed to Gregorio Fontana and was used by eighteenth-century Italian writers. The term appeared in English in George Peacock's 1816 translation of Lacroix's *Differential and Integral Calculus*.

Let's do one more problem on conversion of the polar equation into the rectangular form. This problem demands some creativity and not-so-obvious steps.

Sample Problem 8

Convert equation $r = 4\sin\theta$ to rectangular form.

Step 1: Use the following conversion formula:

$$x^2 + y^2 = r^2$$

Step 2: Use the original polar equation $r = 4\sin\theta$ and one of the conversion formulas, $y = r\sin\theta$. From the conversion formula you obtain:

$$\sin\theta = \frac{y}{r}$$

Step 3: Replace $\sin\theta$ in the original polar equation with the right side of the last expression to obtain:

$$r = 4\sin\theta = 4 \cdot \frac{y}{r} = \frac{4y}{r}$$

$$r = \frac{4y}{r} \text{ or } r^2 = 4y$$

Step 4: Plug in the right side of the last expression instead of r^2 in the expression in Step 1:

$$x^2 + y^2 = r^2$$

$$x^2 + y^2 = 4y \text{ or } x^2 + y^2 - 4y = 0$$

You obtain the rectangular form of the polar equation $r = 4\sin\theta$.

Solution: The rectangular form is $x^2 + y^2 - 4y = 0$.

The previous example proves that converting a polar equation to rectangular form usually requires considerable ingenuity and mathematical intuition.

Graphs of Polar Equations

In previous chapters, you learned how to sketch the graphs of trigonometric functions in rectangular coordinates. You used the key point method at first. It was later augmented by considering periods, asymptotes, shifting, stretching, and shrinking. You were also advised to use a graphing calculator to sketch the graphs.

This section also approaches curve sketching in the polar coordinate system by point plotting first. You plot several points (r, θ) that satisfy the equation and then draw a smooth curve through them.

Sample Problem 9

Sketch the polar graph of $r = 2\cos\theta$.

Step 1: Make a table of values to find values of r for selected values of θ.

θ	0°	30°	45°	60°	90°	120°	135°	150°	180°
r	2	$\sqrt{3}$	$\sqrt{2}$	1	0	−1	$-\sqrt{2}$	$-\sqrt{3}$	−2

θ	210°	225°	240°	270°	300°	315°	330°	360°
r	$-\sqrt{3}$	$-\sqrt{2}$	−1	0	1	$\sqrt{2}$	$\sqrt{3}$	2

Step 2: Sketch the graph. From the table you can see that the graph is complete for values of the angle between 0° and 180°. For angles between 180° and 360°, the same graph is traced over again.

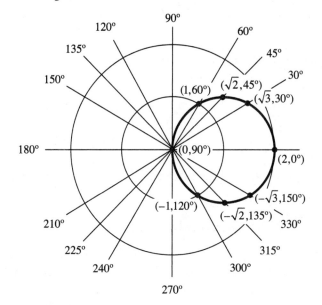

The corresponding figure suggests that the graph in the polar coordinates is a circle.

Solution: The graph of $r = 2\cos\theta$ is a circle.

Similarly, you can sketch a graph for a polar equation $r = 2\sin 2\theta$. The resulting graph is shown on the following figure.

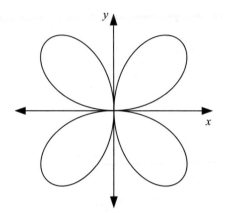

This curve is called a rose curve with four petals. The corresponding rectangular equation is $(x^2 + y^2)^3 = 16x^2y^2$. Looking at this equation, you can predict that sketching it, especially without help from a graphing calculator or special software, would be extremely challenging.

The polar equations of the following curves are simple, but their corresponding rectangular equations are complicated.

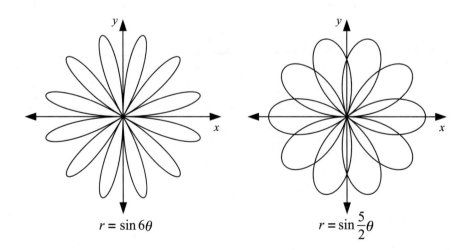

$r = \sin 6\theta$ $r = \sin \dfrac{5}{2}\theta$

Generally, most curves of polar equations are easy to sketch using a graphing calculator. You can read more on this in Chapter 22.

Practice Problems

Problem 1: Find three additional polar representations of the point $\left(2, \dfrac{3\pi}{4}\right)$, using $-2\pi < \theta < 2\pi$.

Problem 2: Find the corresponding rectangular coordinates for the point $\left(-1, \dfrac{5\pi}{4}\right)$.

Problem 3: Find the corresponding polar coordinates for the point $\left(\dfrac{1}{2}, -\dfrac{\sqrt{3}}{2}\right)$.

Problem 4: Find the corresponding polar coordinates for the point $(-3,4)$.

Problem 5: Convert the polar equation $r = 1 - \sin\theta$ to rectangular form.

The Least You Need to Know

- For rectangular coordinates, each point (x, y) has a unique representation, whereas the same point in polar coordinates can be represented by an infinite number of polar coordinates.
- A point in a polar system can be represented as $(r, \theta \pm 2n\pi)$ or $(-r, \theta \pm (2n + 1)\pi)$, where n is any integer.
- The conversion formulas between the two coordinate systems are the following:

    $x = r \cos\theta$

    $y = r \sin\theta$

    $\tan\theta = \dfrac{y}{x}$

    $r^2 = x^2 + y^2$

- The polar equations of rose curves and other important curves are simple, whereas the corresponding rectangular equations are difficult.

Complex Numbers and Operations with Them

In This Chapter

- Introducing the complex plane
- Multiplying complex numbers in polar form
- Finding the powers of complex numbers
- Using a formula to find the roots of a complex number

In Chapter 3, we posed a question about whether complex numbers reside on the number line and we promised to return to this question later in the book.

As you have probably guessed, there is no place on the number line for complex numbers because only real numbers live there. That is why this line sometimes is called the real number line. Well, then, where are the complex numbers located? Can you even represent them geometrically on the plane? In this chapter, we discuss this question and other things about complex numbers.

Geometric Representation of Complex Numbers

At the beginning of the nineteenth century, Jean Robert Argand, a Parisian bookkeeper, published his work where he first presented complex numbers geometrically. He used the so-called complex numbers plane, or simply a *complex plane*. One axis of this plane is the *real axis*, and a typical point on this line has coordinates $(a,0)$ and represents the complex number $a + 0i$. Another axis, a vertical one, is called the *imaginary axis*. A point on it has coordinates $(0,b)$ and represents $0 + bi$—a pure imaginary number that can be either positive or negative.

DEFINITION

A complex number plane features a horizontal axis with real number values, called the **real axis,** and a vertical axis with imaginary numbers, called the **imaginary axis.**

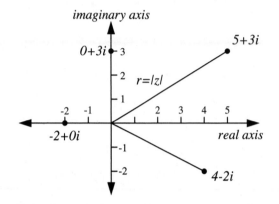

The point representing the complex number $z = a + bi$ can be given either in rectangular coordinates (a,b) or in polar coordinates (r,θ) as in the following figure. Then, you can identify a complex number in two ways:

Rectangular form: $z = a + bi$

Polar form: $z = r\cos\theta + (r\sin\theta)i$

$z = r(\cos\theta + i\sin\theta)$

$z = rcis\theta$

As you can see, the polar form can be abbreviated as $rcis\theta$, which is pronounced "r siss theta." For example, you write $3cis45°$ as the abbreviation for $3(\cos45° + i\sin45°)$.

Sometimes the polar form is also called the trigonometric form of a complex number.

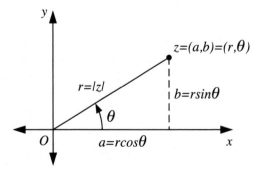

The length of the arrow representing z is called the *absolute value of a complex number* z. Because $r = \sqrt{a^2 + b^2} = |z|$, which is nonnegative, you see only positive values of r in polar coordinates of a complex number z. The angle θ is called the polar angle of z. Throughout this chapter, we give answers to problems using only positive values for r and positive polar angles.

> **DEFINITION**
>
> Let $z = a + bi$ be a complex number. Its polar form is $z = r(\cos\theta + i\sin\theta)$, where $a = r\cos\theta$ and $b = r\sin\theta$, $r = \sqrt{a^2 + b^2}$, and $\tan\theta = \dfrac{b}{a}$. The **absolute value of a complex number z** is defined as $|z| = r = \sqrt{a^2 + b^2}$.

Converting from One Form to Another

In this section, you convert complex numbers from rectangular form to polar form and vice versa. Many graphing calculators are capable of doing this conversion easily. Read about this function of the graphing calculators in Chapter 22.

Sample Problem 1

Express $9cis\dfrac{4\pi}{3}$ in rectangular form.

Step 1: Use the formulas:

$$a = r\cos\theta$$
$$b = r\sin\theta$$

Step 2: Then the rectangular form is:

$$z = a + bi$$
$$a + bi = r\cos\frac{4\pi}{3} + r\sin\frac{4\pi}{3}i$$
$$z = 9cis\frac{4\pi}{3} = 9\cos\frac{4\pi}{3} + 9\sin\frac{4\pi}{3}i$$

Step 3: Substitute values for r and for the sine and cosine of the angle:

$$a + bi = 9\left(-\frac{1}{2}\right) + 9\left(-\frac{\sqrt{3}}{2}\right)i = -\frac{9}{2} - \frac{9\sqrt{3}}{2}i$$

Solution: The rectangular form is $-\frac{9}{2} - \frac{9\sqrt{3}}{2}i$.

If the exact values for the sine and cosine of the polar angle are not available, find approximate values using either the trigonometric tables or a graphing calculator. Subsequently, the rectangular form will be approximate as well, as the next sample problem illustrates.

Sample Problem 2

Express $12cis100°$ in rectangular form.

Step 1: Use the formulas:

$$a = r\cos\theta$$

$$b = r\sin\theta$$

Step 2: Then the rectangular form is:

$$z = a + bi = 12cis100° = 12\cos100° + 12\sin100°i$$

Step 3: Substitute values for r and for the sine and cosine of the angle:

$$a + bi \approx 12(-0.1736) + 12(0.9848)i \approx -2.08 + 11.82i$$

Solution: The rectangular form is $-2.08 + 11.82i$.

Converting from rectangular form to polar form is straightforward as well, as the next problem illustrates.

Sample Problem 3

Express $-2 - 4i$ in polar form.

Step 1: Find r. By definition of the absolute value:

$$r = |z| = \sqrt{a^2 + b^2} = \sqrt{(-2)^2 + (-4)^2} = \sqrt{4 + 16} = \sqrt{20} = 2\sqrt{5}$$

Step 2: Find the angle using the fact that the tangent of the angle is:

$$\tan\theta = \frac{b}{a} = \left(\frac{-4}{-2}\right) = 2$$

Find the angle: $\theta = \tan^{-1}(2) \approx 63°$.

Step 3: Because the complex number is located in the third quadrant (negative a and negative b), you need to add 180° to the obtained angle to get the third-quadrant angle:

$$63° + 180° = 243°$$

Thus, you find that the polar angle is 243°.

Step 4: The polar form of $-2 - 4i$ can be written as:

$$2\sqrt{5}\,cis\,243°$$

Solution: The polar form is $2\sqrt{5}\,cis\,243°$.

Product of Two Complex Numbers in Polar Form

In Chapter 3, you learned how to multiply two complex numbers in rectangular form. In this section, you learn how to multiply two complex numbers in polar form. The following problem illustrates the application of this rule.

WORD OF ADVICE

To multiply two complex numbers in polar form, multiply their absolute values and add their polar angles.

If $z_1 = r_1\,cis\,\alpha$ and $z_2 = r_2\,cis\,\beta$, then: $z_1z_2 = (r_1\,cis\,\alpha)(r_2\,cis\,\beta) = r_1r_2\,cis(\alpha + \beta)$.

Sample Problem 4

Express $z_1 = 3 + 3i$, $z_2 = 2 + 2i\sqrt{3}$, and z_1z_2 in polar form.

Step 1: To multiply numbers in polar form, you need to convert each to the polar form from the rectangular form. Let's convert the first one. Find its absolute value:

$$r_1 = |z_1| = \sqrt{3^2 + 3^2} = \sqrt{18} = \sqrt{9 \cdot 2} = 3\sqrt{2}$$

Find the angle:

$$\tan^{-1}\left(\frac{b}{a}\right) = \tan^{-1}\left(\frac{3}{3}\right) = \tan^{-1}(1) = 45°$$

Therefore, the first complex number in polar form is $3\sqrt{2}cis45°$.

Step 2: Convert the second complex number to polar form. First, find its absolute value:

$$r_2 = |z_2| = |2 + 2i\sqrt{3}| = \sqrt{2^2 + 2^2 \cdot \sqrt{3}^2} = \sqrt{16} = 4$$

Find the angle:

$$\tan^{-1}\left(\frac{b}{a}\right) = \tan^{-1}\left(\frac{2\sqrt{3}}{2}\right) = \tan^{-1}(\sqrt{3}) = 60°$$

Therefore, the second complex number in polar form is $4cis\ 60°$.

Step 3: Multiply two complex numbers using the rule:

$$z_1z_2 = r_1r_2cis\ (\alpha + \beta)$$

Plug in the values for the angles and the absolute values:

$$z_1 \cdot z_2 = \left(3\sqrt{2} \cdot 4\right)cis\left(45° + 60°\right) = 12\sqrt{2}cis105°$$

Solution: The polar forms are $z_1 = 3\sqrt{2}cis45°$, $z_2 = 4cis60°$, and $z_1z_2 = 12\sqrt{2}cis105°$.

Let's do one more problem and use the same rule to find the product of two complex numbers.

Sample Problem 5

Express the product $\left(8cis\frac{\pi}{3}\right)\left(\frac{1}{2}cis\left(-\frac{2\pi}{3}\right)\right)$ in polar and rectangular form.

Step 1: Use the formula for the product:

$$z_1z_2 = r_1r_2cis(\alpha + \beta)$$

$$z_1z_2 = \left(8cis\frac{\pi}{3}\right)\left(\frac{1}{2}cis\left(-\frac{2\pi}{3}\right)\right) = \left(8 \cdot \frac{1}{2}\right)cis\left(\frac{\pi}{3} + \left(-\frac{2\pi}{3}\right)\right) = 4cis\left(-\frac{\pi}{3}\right) = 4cis\left(\frac{5\pi}{3}\right)$$

Note that you substitute a negative value with a positive value for the angle because you use only positive values for radii and angles.

Step 2: Convert product to rectangular form using the values for the sine and cosine of the angle:

$$z_1 z_2 = r cis\theta = r\left(\cos\theta + i\sin\theta\right) = a + bi$$

$$z_1 z_2 = 4cis\left(\frac{5\pi}{3}\right) = 4\left(\cos\frac{5\pi}{3} + i\sin\frac{5\pi}{3}\right) = 4\left(\frac{1}{2} + i\left(-\frac{\sqrt{3}}{2}\right)\right) = 2 - 2i\sqrt{3}$$

Solution: The product in polar form is $4cis\left(\dfrac{5\pi}{3}\right)$ and in rectangular form, $2 - 2i\sqrt{3}$.

De Moivre's Theorem and Powers of Complex Numbers

The polar form of complex numbers enables you to find powers of complex numbers. To find the square of a complex number, you use the rules for multiplication that were discussed previously. For example:

$$(r\ cis\ \alpha)^2 = (r\ cis\ \alpha)(r\ cis\ \alpha) = r \cdot r \cdot cis(\alpha + \alpha) = r^2 cis2\alpha$$

To find the cube of a complex number, use the square of the number and the same rule:

$$(r\ cis\ \alpha)^3 = (r\ cis\ \alpha)^2(r\ cis\ \alpha) = (r\ cis\ 2\alpha)(r\ cis\ \alpha) = r^2 \cdot r \cdot cis(2\alpha + \alpha) = r^3 cis3\alpha$$

You can generalize the previous pattern to obtain a general formula for powers of complex numbers. This formula is called *De Moivre's Theorem*.

DEFINITION

If $z = r\ cis\ \theta$, then $z^n = r^n\ cis\ n\ \theta$, where n is any positive integer. This is called **De Moivre's Theorem.**

By the way, the equation for De Moivre's Theorem holds true for any integer and even for fractional values of n.

Let's use De Moivre's Theorem to solve some problems.

Sample Problem 6

If $z = \dfrac{1}{2} + \dfrac{\sqrt{3}}{2}i$, find z^2, z^3, z^4, z^5, and z^6. Plot these on the complex plane.

Step 1: Convert z to the polar form.

First, find the absolute value:

$$r = |z| = \sqrt{\left(\frac{1}{2}\right)^2 + \left(\frac{\sqrt{3}}{2}\right)^2} = 1$$

Then, find the angle:

$$\tan^{-1}\left(\frac{\frac{\sqrt{3}}{2}}{\frac{1}{2}}\right) = \tan^{-1}\left(\sqrt{3}\right) = 60°$$

Therefore, the polar form of z is $1 cis 60°$.

Step 2: Use De Moivre's Theorem to find the powers of complex number z:

$$z^2 = 1^2 cis\left(2 \cdot 60°\right) = 1 cis 120° = -\frac{1}{2} + \frac{\sqrt{3}}{2}i$$

$$z^3 = 1^3 cis\left(3 \cdot 60°\right) = 1 cis 180° = -1$$

$$z^4 = 1^4 cis\left(4 \cdot 60°\right) = 1 cis 240° = -\frac{1}{2} - \frac{\sqrt{3}}{2}i$$

$$z^5 = 1^5 cis\left(5 \cdot 60°\right) = 1 cis 300° = \frac{1}{2} - \frac{\sqrt{3}}{2}i$$

$$z^6 = 1^6 cis\left(6 \cdot 60°\right) = 1 cis 360° = 1$$

Step 3: Plot the six powers of z on the complex plane:

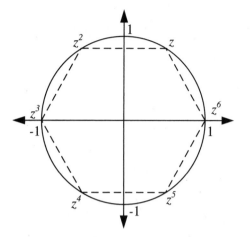

You can see from the illustration that these six powers of z are the vertexes of a regular hexagon inscribed in a circle of radius 1.

Solution: The powers of z are $-\frac{1}{2}+\frac{\sqrt{3}}{2}i$, -1, $-\frac{1}{2}-\frac{\sqrt{3}}{2}i$, $\frac{1}{2}-\frac{\sqrt{3}}{2}i$, and 1.

The path taken by the points corresponding to the powers of a given complex number z depends on the value of z. Some of the paths present equiangular spirals. These curves are called *equiangular spirals* because they cross all lines through the origin at the same angle. These spirals can be found in many natural phenomena such as the horns and claws of some animals, the seeds of a sunflower, and so forth. Let's do another sample problem and sketch this curve.

DEFINITION

Equiangular spirals are formally defined as spiral curves that cut all radial lines at a constant angle. They are also sometimes called logarithmic spirals, Bernoulli spirals, or logistiques.

Sample Problem 7

If $z=\frac{1}{2}+\frac{1}{2}i$, find z^2, z^3, z^4, z^5, z^6, z^7, and z^8. Plot these on the complex plane.

Step 1: Convert z to the polar form.

Find the absolute value first:

$$r = |z| = \sqrt{\left(\frac{1}{2}\right)^2 + \left(\frac{1}{2}\right)^2} = \sqrt{\frac{1}{2}} = \frac{\sqrt{2}}{2}$$

Find the angle:

$$\tan^{-1}\left(\frac{\frac{1}{2}}{\frac{1}{2}}\right) = \tan^{-1}(1) = 45°$$

Therefore, the polar form of z is $\frac{\sqrt{2}}{2}\,cis\,45°$.

Step 2: Use De Moivre's Theorem to find the powers of complex number z:

$$z^2 = \left(\frac{\sqrt{2}}{2}\right)^2 cis(2 \cdot 45°) = \frac{2}{4}cis\,90° = \frac{1}{2}(\cos 90° + i\sin 90°) = \frac{1}{2}i$$

$$z^3 = \left(\frac{\sqrt{2}}{2}\right)^3 cis(3 \cdot 45°) = \frac{2\sqrt{2}}{8}cis\,135° = \frac{\sqrt{2}}{4}(\cos 135° + i\sin 135°) =$$

$$\frac{\sqrt{2}}{4}\left(-\frac{\sqrt{2}}{2} + i\frac{\sqrt{2}}{2}\right) = -\frac{1}{4} + \frac{1}{4}i$$

$$z^4 = \left(\frac{\sqrt{2}}{2}\right)^4 cis(4 \cdot 45°) = \frac{4}{16}cis\,180° = \frac{1}{4}(\cos 180° + i\sin 180°) = -\frac{1}{4}$$

$$z^5 = \left(\frac{\sqrt{2}}{2}\right)^5 cis(5 \cdot 45°) = \frac{4\sqrt{2}}{32}cis\,225° = \frac{\sqrt{2}}{8}(\cos 225° + i\sin 225°) = -\frac{1}{8} - \frac{1}{8}i$$

$$z^6 = \left(\frac{\sqrt{2}}{2}\right)^6 cis(6 \cdot 45°) = \frac{8}{64}cis\,270° = \frac{1}{8}(\cos 270° + i\sin 270°) = -\frac{1}{8}i$$

$$z^7 = \left(\frac{\sqrt{2}}{2}\right)^7 cis(7 \cdot 45°) = \frac{8\sqrt{2}}{128}cis\,315° = \frac{\sqrt{2}}{16}(\cos 315° + i\sin 315°) = \frac{1}{16} - \frac{1}{16}i$$

$$z^8 = \left(\frac{\sqrt{2}}{2}\right)^8 cis(8 \cdot 45°) = \frac{16}{256}cis\,360° = \frac{1}{16}(\cos 360° + i\sin 360°) = \frac{1}{16}$$

Step 3: Plot the powers of z on the complex plane:

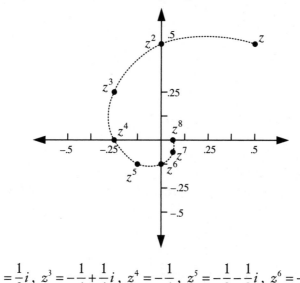

Solution: $z^2 = \dfrac{1}{2}i$, $z^3 = -\dfrac{1}{4} + \dfrac{1}{4}i$, $z^4 = -\dfrac{1}{4}$, $z^5 = -\dfrac{1}{8} - \dfrac{1}{8}i$, $z^6 = -\dfrac{1}{8}i$, $z^7 = \dfrac{1}{16} - \dfrac{1}{16}i$, and $z^8 = \dfrac{1}{16}$

Roots of Complex Numbers

In this section, we use De Moivre's Theorem to find roots of complex numbers. In general, a nonzero complex number can have two square roots, three cube roots, four fourth roots, and k kth roots. Before we provide a general formula for the kth root of a complex number, let's do a sample problem that illustrates the reasoning behind the formula.

Sample Problem 8

Find the cube roots of $8i$.

Step 1: Let's agree that a complex number of the form $z = rcis\alpha$ is a cube root of $8i$. In polar form, a complex number $8i$ can be expressed as $8cis90°$. The angle is equal to $90°$ because the number $0 + 8i$ lies completely on the imaginary axis in the positive direction from the origin.

WORD OF ADVICE

Note that if $a = 0$, then $\theta = \dfrac{\pi}{2} = 90°$ even though $\tan\theta$ is not defined.

Step 2: Because z is a cube root of $8i$, you have:

$z^3 = 8i$ or $(rcis\alpha)^3 = 8i = 8cis90°$

Use De Moivre's Theorem on the left side:

$r^3cis3\alpha = 8cis90°$

Step 3: By looking at both sides of the previous expression, you can conclude that:

$r^3 = 8$ or $r = 2$

Also, $cis3\alpha = cis90°$ or $(90° +$ a multiple of $360°)$ because the values of the sine and cosine will be the same for all these angles.

Therefore, $cis\alpha = cis30°$ or $(30° +$ a multiple of $120°)$. We are utilizing multiples of $120°$ since that measure is a third of $360°$.

Thus, $\alpha = 30°$, $30° + 120°$, $30° + 240°$ or $\alpha = 30°$, $150°$, and $270°$. Because you need three cube roots, you calculate only three angles.

Step 4: Substitute the values for the angles' sines and cosines. Therefore, the three cube roots of $8i$ are:

$$z_1 = 2cis30° = 2\left(\cos 30° + i\sin 30°\right) = 2\left(\frac{\sqrt{3}}{2} + i\frac{1}{2}\right) = \sqrt{3} + i$$

$$z_2 = 2cis150° = 2\left(\cos 150° + i\sin 150°\right) = 2\left(-\frac{\sqrt{3}}{2} + i\frac{1}{2}\right) = -\sqrt{3} + i$$

$$z_3 = 2cis270° = 2\left(\cos 270° + i\sin 270°\right) = 2\left(0 + i(-1)\right) = -2i$$

Solution: The three cube roots of $8i$ are $\sqrt{3} + i$, $-\sqrt{3} + i$, and $-2i$.

Recall that, according to algebra rules about fractional exponents, $x^{\frac{1}{n}} = \sqrt[n]{x}$. For example, $8^{\frac{1}{3}} = \sqrt[3]{8} = 2$, $32^{\frac{1}{5}} = \sqrt[5]{32} = 2$, and so on.

If you closely analyze the cube roots of $8i$ in polar form from the previous problem, you can state that:

$$\sqrt[3]{8i} = \sqrt[3]{8cis\,90°} = \left(8cis\,90°\right)^{\frac{1}{3}}$$

Compare the right side of the preceding expression with the results from the previous problem.

$$\left(8cis\,90°\right)^{\frac{1}{3}} = \left(8\right)^{\frac{1}{3}}\left(cis\,90°\right)^{\frac{1}{3}} = 2cis\left(30° + \frac{k \cdot 360}{3}\right) \text{ for } k = 0,\,1,\,2$$

This leads to the conclusion that to find the cube roots of complex numbers, you need to extract the cube root of the absolute value and divide the angle by 3. You also need to add multiples of $120°(360° \div 3 = 120°)$ to include all three cube roots. You can generalize this pattern to obtain the rule of finding nth roots of a complex number.

> **WORTH KNOWING**
>
> The nth roots of $z = r\,cis\,\theta$ are:
>
> $$\sqrt[n]{z} = z^{\frac{1}{n}} = r^{\frac{1}{n}}cis\left(\frac{\theta}{n} + \frac{k \cdot 360°}{n}\right) \text{ for } k = 0, 1, 2 \ldots n-1$$

Use this rule to solve the last sample problem in this chapter.

Sample Problem 9

Find the cube roots of $1 - i$.

Step 1: Convert $1 - i$ to polar form. Find the absolute value:

$$r = |z| = \sqrt{1^2 + (-1)^2} = \sqrt{2}$$

Find the angle by keeping in mind that the complex number is located in the fourth quadrant, thus the angle is also from the fourth quadrant:

$$\tan^{-1}\left(\frac{-1}{1}\right) = \tan^{-1}(-1) = 315°$$

Then, the polar form of $1 - i$ is $z = \sqrt{2}cis\,315°$.

Step 2: Find the roots using the rule:

$$\sqrt[n]{z} = z^{\frac{1}{n}} = r^{\frac{1}{n}} cis\left(\frac{\theta}{n} + \frac{k \cdot 360°}{n}\right)$$

$$\sqrt[3]{z} = \left(\sqrt{2} cis 315°\right)^{\frac{1}{3}} = \sqrt{2}^{\frac{1}{3}} cis\left(\frac{315°}{3} + \frac{k \cdot 360°}{3}\right)$$

WORD OF ADVICE

Because $\sqrt[3]{\sqrt{2}} = \left(\sqrt{2}\right)^{\frac{1}{3}} = \left(2^{\frac{1}{2}}\right)^{\frac{1}{3}} = 2^{\frac{1}{2}\cdot\frac{1}{3}} = 2^{\frac{1}{6}} = \sqrt[6]{2}$, the general rule is $\sqrt[n]{\sqrt[k]{x}} = \sqrt[nk]{x}$.

For example, $\sqrt[4]{\sqrt[2]{256}} = \sqrt[4\cdot2]{256} = \sqrt[8]{256} = 2$ since $2^8 = 256$.

Note that it is common not to use 2 to indicate the square root: $\sqrt[2]{16} = \sqrt{16}$.

Then, when $k = 0$, the first root is:

$$\sqrt{2}^{\frac{1}{3}} cis\left(\frac{315°}{3} + \frac{0 \cdot 360°}{3}\right) = \sqrt[3]{\sqrt{2}} cis 105° = \sqrt[6]{2} cis 105°$$

When $k = 1$, the second root is:

$$\sqrt{2}^{\frac{1}{3}} cis\left(\frac{315°}{3} + \frac{k \cdot 360°}{3}\right) = \sqrt[6]{2} cis\left(105° + \frac{1 \cdot 360°}{3}\right) = \sqrt[6]{2} cis\left(105° + 120°\right) = \sqrt[6]{2} cis 225°$$

When $k = 2$, the third root is:

$$\sqrt{2}^{\frac{1}{3}} cis\left(\frac{315°}{3} + \frac{k \cdot 360°}{3}\right) = \sqrt[6]{2} cis\left(105° + \frac{2 \cdot 360°}{3}\right) = \sqrt[6]{2} cis\left(105° + 240°\right) = \sqrt[6]{2} cis 345°.$$

Solution: The three cube roots of $1 - i$ are $\sqrt[6]{2} cis 105°$, $\sqrt[6]{2} cis 225°$, and $\sqrt[6]{2} cis 345°$.

The formula for the *n*th roots of a complex number has an interesting geometric interpretation as shown in the following illustration.

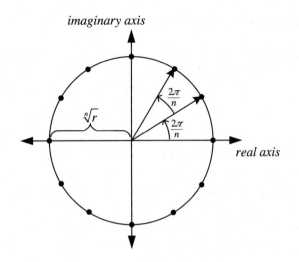

Because the nth roots of z all have the same magnitude $\sqrt[n]{r}$, they all lie on a circle of radius $\sqrt[n]{r}$, with the center at the origin. To add additional mathematical beauty to this fact, note that the n roots are equally spaced along the circle, because nth roots have arguments (angles) that differ by $\dfrac{2\pi}{n}$.

Practice Problems

Problem 1: Express $4cis45°$ in rectangular form.

Problem 2: Express $1+i\sqrt{3}$ in polar form.

Problem 3: Find the product of $3cis\dfrac{\pi}{3}$ and $4cis\dfrac{\pi}{6}$.

Problem 4: Show that $\left(-1+\sqrt{3}i\right)^{24}$ is a real number.

Problem 5: Find the four fourth roots of -16.

The Least You Need to Know

- The point representing the complex number $z = a + bi$ on a complex plane can be given either in rectangular coordinates (a,b) or in polar coordinates (r, θ).

- The absolute value of a complex number z is defined to be $|z| = r = \sqrt{a^2 + b^2}$.

- To multiply two complex numbers in polar form, multiply their absolute values and add their polar angles.

- To find powers of complex numbers, use De Moivre's Theorem: if $z = r$ cis θ, then $z^n = r^n$ cis $n\,\theta$, where n is any positive integer.

- The nth roots of $z = r$ cis θ are as follows:

$$\sqrt[n]{z} = z^{\frac{1}{n}} = r^{\frac{1}{n}} \mathrm{cis}\left(\frac{\theta}{n} + \frac{k \cdot 360°}{n}\right) \text{ for } k = 0, 1, 2 \dots n - 1$$

Trigonometry and Calculators

In This Chapter

- Introducing the arsenal of electronic helpers
- Using a calculator with trigonometric functions
- Evaluating inverse trigonometric functions with calculators
- Sketching graphs of trig functions with calculators
- Solving trig equations with calculators

In the preceding chapters, we mentioned the great help that a scientific or graphing calculator can be. Your electronic friend is not only an efficient tool to evaluate regular and inverse trigonometric functions, but it also helps you solve trigonometric equations and graph trigonometric functions.

First, we discuss the specific types of widely used calculators, as well as programs on computers and other handheld devices. Then, we illustrate the specific methods of working with trigonometry on a calculator. We conclude with a guide on how to solve trigonometric equations, both analytically and graphically.

Calculators That Can Help You with Trigonometry

As recently as a few years ago, the arsenal of calculators commonly used by students studying trigonometry was rather sparse. Sure, there were a variety of specific models of calculators made by different companies, but all of them fit into three main types: *standard calculators*, *scientific calculators*, and *graphing calculators*. These were available as dedicated handheld electronic devices with the fancier ones costing quite a bit.

Although computers equipped with software such as MathLab or Mathematica were also available, it was rare for the average student to use them in the daily study of trigonometry due to the lack of convenience.

DEFINITION

A **standard calculator** usually has just the basic arithmetic functions: addition, subtraction, multiplication, and division.

A **scientific calculator** is a type of calculator designed to calculate problems in science and mathematics. They have many advanced mathematical functions, including trigonometric ones: sine, cosine, tangent, arcsine, arccosine, and arctangent.

A **graphing calculator** is a type of calculator capable of plotting graphs, solving simultaneous equations, and performing numerous other tasks with variables.

A variety of other devices that can potentially help you with trigonometry have reached the market. Because modern electronic organizers, cell phones, and tablets are on par in terms of computational power with computers of previous generations, it did not take long for software calculators to be written for these devices. For example, many popular applications for iPhone, Android, and Blackberry cellular phones do a good job of serving as scientific and graphing calculators.

An easily utilized tool that's often overlooked is the Internet itself. Search engines such as Google are often able to understand and evaluate trigonometric expressions, even complex ones with multiple sets of nested parentheses. Likewise, free tools like Wolfram Alpha give you the capabilities of powerful mathematics software (which serves as a graphing calculator and some more) anywhere with Internet access, be it a library, your friend's house, or on a long plane ride with a tablet in your hands.

DANGEROUS TURN

As tempting as using a calculator is, keep in mind that using it continuously to solve problems directly versus actually doing the math is not going to help you learn and practice the trigonometric concepts that this book is about. Try to do all the intermediate steps without a calculator and save it for cases when you need to get the final numerical solution.

Considering such a large variety of tools at your disposal, this chapter seeks to give you the key methods and guidelines of how to properly use your electronic helpers. The exact type of calculator or device you use is up to you.

Evaluating Trigonometric Functions

After following extensive proofs and algebraic solution steps in the previous chapters of this book, you must be relieved that you have gotten to the simple part. Calculators do indeed make it easy to evaluate trigonometric functions. However, you do need to keep in mind a few key trigonometric concepts.

Most calculators default their trigonometry calculators in radians. In order to perform calculations in degrees, you must use the two conversion formulas provided in Chapter 8:

$$1 \text{ radian} = \frac{180}{\pi} \text{ degrees} \approx 57.2957 \text{ degrees}$$

$$1 \text{ degree} = \frac{\pi}{180} \text{ radians} \approx 0.0174533 \text{ radians}$$

Some calculators make life easier by having a setting that can toggle between calculations in radian and degree modes. This can be accomplished either via a button on the front of the calculator (common among scientific calculators) or a change in the calculator settings (common among graphing calculators). Some software calculators, such as the one built into the Google search engine, can also understand your input if you specifically state whether you want it to think in radians or degrees. For example, typing "sin(25)" or "sin(25 radians)" into Google displays a result of –0.13235175, equal to the sine of 25 radians. Meanwhile, typing "sin(25 degrees)" produces 0.422618262, which is equal to the sine of 25 degrees.

DANGEROUS TURN

Always remember that a calculator cannot read your mind regarding whether you are asking it to do a calculation in radians or degrees. Make sure that the option that toggles between degree and radian calculations is set to the correct value. Specific settings aside, most calculators default to performing calculations in radians. If the end result of the calculation does not seem right, be sure to check your work!

Most scientific and graphing calculators have three buttons, one each for the sine, cosine, and tangent function. Scientific calculators generally require that you press the function button first, then type the value of the angle you want to calculate (in the correct unit) and press **Enter**. The calculator will then display the positive or negative value of the function. Keep in mind that the tangent function does not have value for angles of $\frac{\pi}{2} + n\pi$ radians and their degree equivalents. Thus, trying to calculate the tangent of such value will display either a large number (numerically as close to infinity as the calculator has the capacity to get) or an error.

Graphing calculators function in a similar way to the scientific ones with one core difference: some models automatically insert an opening parenthesis between the trigonometric function and the value of the angle. For those calculator models, a closing parenthesis needs to be inserted.

Another key capability of graphing calculators is that they can compute multiple nested expressions. Therefore, you can calculate sines, cosines, and tangents of whole algebraic expressions, such as $\sin\left(\frac{\pi}{2} + \frac{\pi}{4}\right)$, or even trigonometric functions nested within each other, such as $\sin\left(\cos\left(\frac{\pi}{6}\right)\right)$.

Sample Problem 1

Evaluate the following expressions on a scientific calculator (or its software equivalent on a mobile device): $\sin\left(\frac{\pi}{12}\right)$, $\cos\left(\frac{\pi}{6} + \frac{\pi}{3}\right)$, $\tan(37°)$.

Step 1: Start by computing the numerical equivalents of $\sin\left(\frac{\pi}{12}\right)$ and $\cos\left(\frac{\pi}{6} + \frac{\pi}{3}\right)$.

Some calculators have a π key that makes entering in the value of π precise and simple.

If your calculator does not have that key, you can use the approximate value of π to four decimal places: 3.1416.

Thus, you obtain $\frac{\pi}{12} \approx \frac{3.1416}{12} \approx 0.2618$ radians for the first expression and

$\frac{\pi}{6} + \frac{\pi}{3} \approx \frac{3.1416}{6} + \frac{3.1416}{3} \approx 0.5236 + 1.0472 \approx 1.5708$ radians for the second one.

Step 2: Let's evaluate the three expressions to obtain the final results. Remember to have the calculator set to radians for the first two calculations and to degrees for the third.

Calculating in radians: $\sin(0.2618) \approx 0.2588$

Calculating in radians: $\cos(1.5708) \approx 0$ (Your calculator might give you a negative number that is close to zero. This is due to the rounding performed in the previous steps.)

Calculating in degrees: $\tan(37°) \approx 0.7536$

Solution: The three expressions are equal to approximately 0.2588, 0, and 0.7536.

DANGEROUS TURN

Remember that many teachers and professors require you to show algebraic steps in the solutions of assignments and ban scientific or graphing calculators outright on tests and quizzes. Likewise, some standardized tests have also banned calculator use.

In sum, although you can (and should) take advantage of efficiency gains that a good calculator can give you, never lose sight of learning the core principles and methods of trigonometry itself.

Sample Problem 2

Evaluate the following expression on a graphing calculator:

$$\sin\left(28° + \frac{\pi}{4}\right) - \cos\left(\frac{\pi}{6} + \frac{\pi}{2}\right) + \tan(1_{rad}).$$

Step 1: Although graphing calculators can generally evaluate even complex expressions in one calculation step, a closer look at the example shows one complication. The expression contains values in both degrees and radians. In order to tackle it, you need to convert all angle values to a common measure.

Converting everything to either degrees or radians is an option. You can choose to convert the values to radians in this case because all of them except one are already in that format.

Recall the degrees-to-radians conversion formula:

$$1 \text{ radian} = \frac{180}{\pi} \text{ degrees} \approx 57.2957 \text{ degrees}$$

To obtain the angle value in radians, you need to divide the degree value by 57.2957 degrees per radian.

$$\frac{28°}{57.2957} \approx 0.4887 \text{ radians}$$

Step 2: Now that you have all of the angle values in radians, reconstruct the original expression as you need to enter it into a typical graphing calculator. Keep in mind that most of the commonly used calculator models have either a key for π or a function of pi() that calculates it. The latter is used in the following example.

$\sin\left(0.4887_{rad}+\dfrac{\pi}{4}\right)-\cos\left(\dfrac{\pi}{6}+\dfrac{\pi}{2}\right)+\tan\left(1_{rad}\right)$ can thus be entered in to the graphing calculator as "sin(0.4887+pi()/4)-cos(pi()/6+pi()/2)+tan(1)".

Step 3: Press **Enter** and obtain the result of 3.0137.

Solution: The expression is equal to approximately 3.0137.

> **DANGEROUS TURN**
>
> Be careful with entering large expressions into a graphing calculator. Always keep in mind the order of operations—division and multiplication steps are calculated before addition and subtraction. It is also often helpful to include extra sets of parentheses to help ensure that the calculator interprets operations in the same order as intended in the original expression.

Evaluating Inverse Trigonometric Functions

Evaluating inverse trigonometric functions on a calculator is a process similar to evaluating the regular ones. Most calculators have the inverse functions accessible via the same keys as the regular ones, the only difference being that there is a modifier key (oftentimes labeled "2nd") that you are required to press first. Likewise, the keys are usually marked with the labels $\sin^{-1}$, $\cos^{-1}$, and $\tan^{-1}$ right above them.

Similar to the regular trigonometric functions, it is important to know whether the calculator is set in radians or degrees mode. In the case of inverse functions, the result of the calculation will be expressed as a degree measure.

Sample Problem 3

Evaluate the following expressions on a graphing calculator, obtaining results in both radians and degrees: arcsin(.5), arccos(.75), and arctan(50).

Step 1: Let's first perform the calculations for all three expressions in degrees mode.

$\sin^{-1}(.5)= 30°$

$\cos^{-1}(.75)= 41.41°$

$\tan^{-1}(50)= 88.85°$

Step 2: Now, let's switch the calculator to radians mode and perform the same calculations again.

$\sin^{-1}(.5) = 0.5236$ radians

$\cos^{-1}(.75) = 0.7227$ radians

$\tan^{-1}(50) = 1.5508$ radians

Step 3: Although you have already obtained the solution in the previous step, it is a useful exercise to check the two results against each other.

As you might recall from the previous chapters, 30° is equivalent to $\dfrac{\pi}{6}$ radians. You can use your calculator to show that $\dfrac{\pi}{6} \approx 0.5236$ radians. For the second and third results, you can use the conversion formula between radians and degrees to show the equivalence of the answers.

$$41.41° = 41.41° \times \frac{0.0174533 \text{ radians}}{\text{per one degree}} \approx 0.7227 \text{ radians}$$

$$88.85° = 88.85° \times \frac{0.0174533 \text{ radians}}{\text{per one degree}} \approx 1.5508 \text{ radians}$$

Solution: You have obtained the three angles to be equal to 30°, 41.41°, and 88.85° or 0.5236, 0.7227, and 1.5508 radians and have also shown the equivalence of the two sets of answers.

Sketching Graphs of Trigonometric Functions

As evidenced by their name, graphing calculators have the capability not just to evaluate small or large trigonometry expressions but also to graph their values over a specific domain. Although the specific graphing utilities across various calculator models can be different in terms of layout and specific keys to press, they all follow several common design principles.

A typical graphing utility has a section where you input the functions to be graphed, a section with options for the graph, and of course the actual graphing window where the final result appears. The two most important options to consider are the domain and range of the graph. As discussed in the previous chapters, the domain is the set of *x* values for which you evaluate and graph the function whereas the range is the set of *y* values that the function takes.

Calculators generally require that you define a specific viewing domain because that will define the horizontal boundaries of the graphing window. Because many trigonometric functions have a domain all the way from negative infinity to positive infinity, the calculator's window will only show a specific neighborhood of a function. In most cases, the neighborhood you want to analyze is around 0 and includes values on one (or both) sides of it. In order to set a proper viewing domain, it is helpful to know the period of the function and to set a window size so that several periods of the function can be shown.

The domain for the y-axis of the window can also be defined manually or, in many calculator models, can be evaluated automatically based on the function values over the x-range. In the case of automatic evaluation, the calculator evaluates all the function values over the entire x-range and chooses a y-range so that all of these values are displayed.

Together the x- and the y-ranges determine what is commonly known as the *viewing rectangle*. Proper choice of these ranges is essential because there is not much use in using the graphing capability if you cannot see the results. If you try to graph a function and it does not show up (or only a small part of it shows up), always go back to the window options and choose a different domain and range, preferably slightly larger than the area of the function that you are trying to analyze. Many calculator models also offer zoom in and zoom out functionality to make setting the domain faster and easier.

DEFINITION

A **viewing rectangle** is the size of the window on the graphing calculator that is defined by a range of x- and y-values that are visible when the calculator graphs a function.

Sample Problem 4

Plot the graph of $y = 3\sin2x + 2$ using a graphing calculator.

Step 1: Let's first determine the appropriate viewing rectangle before graphing the function. From the previous chapters, you already know that the period of the $y = \sin x$ function is 2π. Thus, a function of $\sin2x$ will have a period twice as small, which will be equal to π. To capture a complete period of the function, you need to set a viewing window with an x-range starting slightly below 0 and ending just above π.

Step 2: The range of the viewing window can be determined either automatically (if your calculator allows it) or manually. Let's try the manual approach. As you might recall, the range of the regular $y = \sin x$ function is from –1 to 1. In this case, the sin x function is multiplied by 3, which would make the effective range from –3 to 3. Keep in mind that 2 is also added at the end, which shifts the entire range to be from –1 to 5.

Step 3: Taking the domain and the range from the previous steps, let's graph the function $y = 3\sin 2x + 2$ on the domain from –0.5 to 3.5 and on the range from –1 to 5. The result is shown in the following graph and should match the one you obtained for a similar problem in Chapter 8.

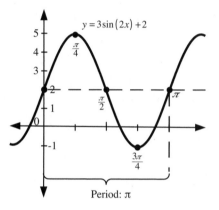

Solution: You have graphed the function $y = 3\sin 2x + 2$ using a graphing calculator.

Although graphing the cosine function is similar to graphing the sine, the tangent function does present some interesting challenges in using the graphing calculator. First of all, the y-range of the tangent function is from negative infinity to positive infinity. Hence, the calculator will have difficulty automatically calculating it and you will need to simply choose a reasonable value in order to see the behavior of the function.

The other difference in the tangent function is the fact that its values jump from positive to negative infinity. Graphing calculators generally don't show this disappearance of the graph and show the continuous plot, which often features a vertical line where that jump takes place. It is important to note that although this line falls within the place where the asymptote should be, it is not really one. The calculator is simply attempting to connect all of the function values on the graph.

One way to eliminate the vertical lines is to set the calculator to dot mode. (You can refer to your calculator manual to see how to do it for your particular model). In this case, the calculator just plots individual dots and does not attempt to connect them.

Solving Trig Equations with a Calculator

In previous chapters of this book, you learned how to solve trigonometric equations analytically. Recall that solving an equation involves finding the values of x that make both sides equal. The graphing calculator enables you to solve equations graphically by graphing each individual side and seeing where the graphs intersect. The x-values of the intersection points of two lines on the graph are the x-values that make both sides of the equation equal, hence its solutions.

A common way to transform an equation in order to make it easier to solve analytically is to add or subtract terms from both sides to result in one side being equal to zero. This method also comes in handy when solving equations graphically because, when one side of the equations equals zero, all you have to do is find the x-values of the graph where $y = 0$, also known as the *x-intercepts*.

DEFINITION

An *x-intercept* is the value of a function at which it intersects the *x*-axis, which happens at values where $y = 0$.

Keep in mind that just like with analytical methods, solving equations graphically can yield multiple solutions—which would be multiple intersections of the graphs either with each other or, in the case of the expression being equal to zero, multiple x-intercepts.

Likewise, it is also important to keep in mind the periodic nature of trigonometric functions and the periodic nature of solutions of many trigonometry equations. These solutions will be intercepts that repeat regularly on the graph. Similar to analytic methods, you are often interested in finding solutions just within the 0 to 2π range, or a similar appropriate range given the period of the function. This would give you the basic solutions that can then be extended across the entire domain of the function by adding the period of the function to them.

As discussed, the graphing calculator is a useful tool and can greatly speed up and simplify the solutions to many trigonometric problems. Similarly, it can also give you graphical insight into the behavior of many trigonometric functions without having to go through the tedious process of finding values for key points. However, please keep in mind that being able to properly use the calculator requires some basic trigonometric knowledge to begin with and is not a replacement for thorough and complete analytical understanding of the subject.

The Least You Need to Know

- The three main types of calculators are standard calculators, scientific calculators, and graphing calculators.
- Scientific and graphing calculators can calculate the main trigonometric functions: sine, cosine, tangent, arcsine, arccosine, and arctangent.
- Always remember to use your calculator in the correct mode (radians or degrees) and make sure to check your answers to be sure they seem logical.
- For problems involving both radian and degree measures, convert all angle measures into one common unit before evaluating the whole expression on a calculator.
- In graphing trigonometric functions, always make sure to set the appropriate viewing window in order to be able to see the results.
- Solutions to trigonometric equations can be determined graphically by finding intersections between multiple graphs or the x-intercepts of a single graph.

Glossary

absolute value of a complex number The length of the arrow representing the complex number on the complex plane. If $z = a + bi$, then $|z| = \sqrt{a^2 + b^2}$.

acute angle An angle with a measure less than 90°.

amplitude of a periodic function One half of the difference between the maximum and the minimum values of the function.

angle Two rays that share an endpoint.

angle of depression Angle by which an observer's line of sight must be lowered from the horizontal to the point observed.

angle of elevation Angle by which an observer's line of sight must be raised from the horizontal to the point observed.

asymptote A line that a curve approaches closer and closer to a certain x- or y-value but never quite reaches it.

binomial A polynomial that has only two terms.

Cartesian coordinate system Specifies each point uniquely in a plane by a pair of numerical coordinates, which are the signed distances from the point to two fixed perpendicular directed lines, measured with the same unit of length.

central angle An angle with its vertex being the center of the circle.

chord A segment with its endpoints lying on a circle.

cofunction A function g if $f(A) = g(B)$ whenever A and B are complementary angles. Example pairs of cofunctions are sine and cosine, tangent and cotangent, and secant and cosecant.

complex conjugate The conjugate of the complex number $z = a + bi$, which is denoted by $\bar{z} = a - bi$.

complex fraction   A fraction that has other fractions in its numerator, denominator, or both.

complex number   A number in the form $a + bi$, where a and b are real numbers and i is the imaginary unit. a is the real part of such number and bi is the imaginary part.

complex number plane   A plane that features a horizontal axis with real number values and a vertical axis with imaginary numbers.

complementary angles   Two angles with a combined measure of 90°.

concave   Something that curves in or is hollowed inward, as opposed to convex.

conjugate   A binomial with a middle sign opposite of another binomial with the same terms (e.g., $2x + 3$ and $2x - 3$ are conjugates).

convex   Something that curves out or bulges outward, as opposed to concave.

coordinates   An ordered pair of numbers associated with a point on the plane.

cosecant $\csc \theta = \dfrac{r}{y}$, $y \neq 0$, where $P(x,y)$ is a point on a circle with equation $x^2 + y^2 = 1$ and θ is an angle in standard position, which means at the origin, with terminal ray OP. Also, $\csc \theta = \dfrac{1}{\sin \theta}$.

cosine $\cos \theta = \dfrac{x}{r}$, where $P(x,y)$ is a point on a circle with equation $x^2 + y^2 = 1$ and θ is an angle in standard position with terminal ray OP.

cotangent $\cot \theta = \dfrac{x}{y}$, $y \neq 0$, where $P(x,y)$ is a point on a circle with equation $x^2 + y^2 = 1$ and θ is an angle in standard position with terminal ray OP. Also, $\cot \theta = \dfrac{1}{\tan \theta} = \dfrac{\cos \theta}{\sin \theta}$.

coterminal angles   Two angles in standard position that have the same terminal side.

degree   A unit for measuring angles, $\dfrac{1}{360}$ of a revolution.

diameter   A chord that contains the center.

domain of a function   The set of all values for which the function is defined.

equation   A mathematical statement that asserts the equality of two expressions by placing them on two sides of an equals sign.

equilateral triangle   A triangle that has equal sides and three equal angles.

even function A function f such that if $f(-x) = f(x)$.

expression A combination of symbols that are formed according to mathematical rules. An expression does not have an equals sign, unlike an equation.

extreme values of a function The minimum and maximum values of the function.

function A dependent relationship between quantities when every input has exactly one matching output.

imaginary number A number with square that is negative. They have the form bi where b is a nonzero real number and i is the imaginary unit, defined such that $i^2 = -1$.

imaginary unit i The number $\sqrt{-1}$ with the property $i^2 = -1$.

inverse functions Two functions f and g, where $g(f(x)) = x$ for all x in the domain of f and $f(g(x)) = x$ for all x in the domain of g. f^{-1} represents the inverse of f.

isosceles triangle A triangle that has two equal sides and two equal angles.

minor arc of a circle The union of two points on the circle and all the points of the circle that lie in the interior of the central angle whose sides contain the two points.

oblique triangle A triangle that does not have a right angle.

obtuse angle An angle with a measure greater than 90°.

odd function A function f such that if $f(-x) = -f(x)$.

origin The point of intersection of x- and y-axes on a coordinate plane.

period The amount of horizontal space it takes a periodic function to repeat itself.

periodic function A function that repeats its values in regular intervals, which are called periods. The most important examples are the trigonometric functions, which repeat over intervals of length 2π radians.

plane trigonometry The branch of trigonometry that comprises the solution of plane triangles and investigations of plane angles and their functions.

point of inflection A point on a curve at which the sign of the curvature (i.e., the concavity) changes.

polar axis The reference ray, usually the nonnegative x-axis, in the polar coordinate system.

polar coordinates   The ordered pair (r, θ) that describes the position of a point P in the plane.

pole   The origin in the polar coordinate system.

polynomial   The sum or the difference of distinct terms.

radian   A unit for measuring angles, where there are 2π radians in a circle.

radius of a circle   A segment that joins the center of a circle to a point on the circle.

range of a function   The set of all output values of the function.

real numbers   A set of all rational and irrational numbers.

reference angle   The acute angle θ is a reference angle for the angles $180° - \theta$, $180° + \theta$, $360° - \theta$, and all coterminal angles.

relative maximum   A function $f(x)$ has this value at $x = a$ if $f(a)$ is greater than any value on a certain part of its domain. Other values of the function may in fact be greater.

relative minimum   A function $f(x)$ has this value at $x = b$ if $f(b)$ is less than any value in its immediate neighborhood. Other values of the function may in fact be less.

quadrants   The four regions into which the coordinate plane is divided by the x- and y-axes.

quadrilateral   A four-sided polygon.

scalene triangle   A triangle with no equal sides or equal angles.

secant $\sec\theta = \dfrac{r}{x}$, $x \neq 0$, where $P(x, y)$ is a point on a circle with equation $x^2 + y^2 = 1$ and θ is an angle in standard position with terminal ray OP. Also, $\sec\theta = \dfrac{1}{\cos\theta}$.

sine $\sin\theta = \dfrac{y}{r}$, where $P(x, y)$ is a point on a circle with equation $x^2 + y^2 = 1$ and θ is an angle in standard position with terminal ray OP.

spherical trigonometry   The branch of trigonometry concerned with the measurement of the angles and sides of spherical triangles.

supplementary angles   Two angles with a combined measure of $180°$.

tangent $\tan \theta = \dfrac{y}{x}$, $x \neq 0$, where $P(x,y)$ is a point on a circle with equation $x^2 + y^2 = 1$ and θ is an angle in standard position with terminal ray OP. Alternatively, for an acute angle of a right triangle, it can be expressed as the ratio of the length of the leg opposite to the angle to the length of the leg adjacent to the angle.

triangulation The process of determining the location of a point by measuring angles to it from known points at either end of a fixed baseline rather than measuring distances to the point directly. The point can then be fixed as the third point of a triangle with one known side and two known angles.

trigonometric identity An equation that is true for all values of the variable for which each side is defined.

trinomial A polynomial that has three terms. A quadratic trinomial has the form $ax^2 + bx + c$.

unit circle The circle that is described by the equation $x^2 + y^2 = 1$. It is a circle centered on the origin with a radius of 1.

vertical line test Tests whether or not a graph is a function. If any vertical line can be drawn through the graph that intersects the graph more than once, then the graph cannot be a function.

Practice Problems and Answers

Chapter 4 Trigonometric Functions and Right Triangles

Problem 1: Without using the table, find the value of (tan30°)(tan60°).

Using the relations in the 30°-60°-90° right triangle, write: $\tan 30° = \dfrac{\sqrt{3}}{3}$, $\tan 60° = \sqrt{3}$, then $(\tan 30°)(\tan 60°) = \dfrac{\sqrt{3}}{3} \cdot \sqrt{3} = \dfrac{3}{3} = 1$.

Answer: The product is equal to 1.

Problem 2: In a right triangle, the hypotenuse is 17 units long. Find the length of a leg that is adjacent to an acute angle with a measure of 32°.

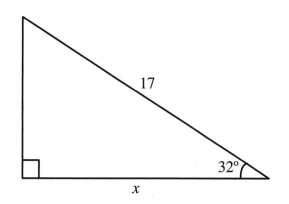

Because we already know the measure of the angle and the length of the hypotenuse, let's set up an equation relating the two and solve it.

$$\cos 32° = \frac{x}{17} \rightarrow x = \cos 32° \cdot 17 = 0.8480 \cdot 17 \approx 14.4$$

Answer: The length of the leg is 14.4 units.

Problem 3: $ABCD$ is an isosceles trapezoid. If BC is 16, AD is 30, and the measure of $\angle C$ is 110°, find AB correct to the nearest whole number.

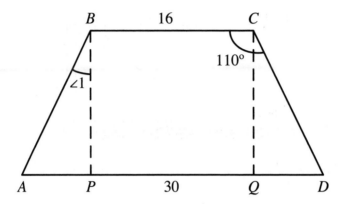

Because the trapezoid is an isosceles one, $AP + QD = 30 - 16 = 14$, then $AP = QD = 7$.

$\angle B = \angle C = 110°$. Then $\angle 1 = 110° - 90° = 20°$; $\sin \angle 1 = \sin 20° = \dfrac{AP}{AB} = \dfrac{7}{AB}$; and

$AB = \dfrac{7}{\sin 20°} = \dfrac{7}{0.3420} \approx 20.$

Answer: AB is 20.

Problem 4: A kite string is 250 meters long. Find the height of the kite if the string makes an angle of 40° with the ground. (Assume the string does not sag.)

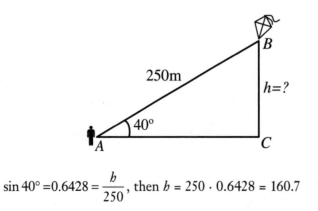

$\sin 40° = 0.6428 = \dfrac{b}{250}$, then $b = 250 \cdot 0.6428 = 160.7$

Answer: The height of the kite from the ground is 160.7 meters.

Problem 5: From the top of a cliff 350 meters high, the angle of depression of a boat measures 26°. How far from the cliff is the ship, to the nearest meter?

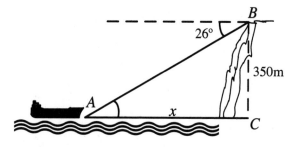

The $\angle ABC$ of the right triangle is $90° - 26° = 64°$. $\tan 64° = 2.0503 = \dfrac{x}{350}$. Then $x = 2.0503 \cdot 350 = 717.605$.

Answer: The ship is approximately 718 meters away from the cliff.

Chapter 5: Relations Among Trigonometric Ratios

Problem 1: Prove that $\cos^2 \theta - \sin^2 \theta = 2\cos^2 \theta - 1$.

Because $\sin^2 \theta + \cos^2 \theta = 1$, $\sin^2 \theta = 1 - \cos^2 \theta$; then the left side becomes $\cos^2 \theta - \left(1 - \cos^2 \theta\right) = \cos^2 \theta - 1 + \cos^2 \theta = 2\cos^2 \theta - 1$.

Answer: You proved that $\cos^2 \theta - \sin^2 \theta = 2\cos^2 \theta - 1$.

Problem 2: Prove that $\sin^2 \theta = \dfrac{1}{\cot^2 \theta + 1}$.

Using reciprocal formulas, write $\sin^2 \theta = \dfrac{1}{\csc^2 \theta}$. Using the third Pythagorean identity, write $\cot^2 \theta + 1 = \csc^2 \theta$. Substitute the left side of the last expression into the right side of the previous one: $\sin^2 \theta = \dfrac{1}{\csc^2 \theta} = \dfrac{1}{\cot^2 \theta + 1}$.

Answer: You proved that $\sin^2 \theta = \dfrac{1}{\cot^2 \theta + 1}$.

Problem 3: The hypotenuse of a right triangle is 12 and one of the acute angles is 37°. Find the other two sides.

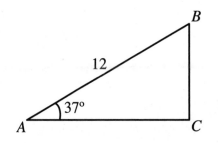

Because $\sin A = \dfrac{BC}{AB} = \dfrac{BC}{12}$, $BC = 12 \cdot \sin 37° = 12 \cdot 0.6018 \approx 7.2$. Similarly, $AC = 12 \cdot \cos 37° = 12 \cdot 0.7986 \approx 9.6$.

Answer: The sides are 7.2 and 9.6.

Problem 4: A rectangle $ABCD$ is 7 centimeters wide and 24 centimeters long. Find the measure of the acute angle between its diagonals.

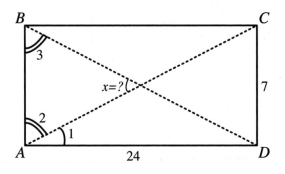

Consider $\triangle ACD$. Because $\tan \angle 1 = \dfrac{7}{24} = 0.2916$, then $m\angle 1 = \tan^{-1}(0.2916) = 16.3°$. Then $\angle 3 = \angle 2 = 90° - 16.3° = 73.7°$. Then $\angle x = 180° - 2 \cdot 73.7° = 180° - 147.4° = 32.6°$.

Answer: The acute angle between the diagonals is 32.6°.

Problem 5: A triangle has sides of lengths 16, 16, and 8. Find the measures of the angles of the triangle to the nearest tenth of a degree.

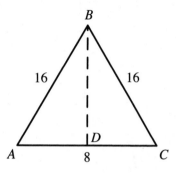

By drawing the altitude to the base, you get two congruent right triangles, where $AD = DC = 4$; then $\cos A = \dfrac{4}{16} = \dfrac{1}{4} = 0.25$. $\angle A = \cos^{-1} 0.25 \approx 75.5°$, thus $\angle A = \angle C \approx 75.5°$ and $\angle B = 180° - 2 \cdot (75.5°) = 29.0°$.

Answer: The angles are approximately 75.5°, 75.5°, and 29.0°.

Chapter 6: The Law of Sines

Problem 1: The area of $\triangle RQP$ is 60. If p is 10, q is 20, and $\angle R$ is an obtuse angle, find the measure of $\angle R$.

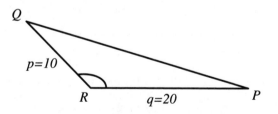

Since the triangle's area is $A = \dfrac{1}{2} pq \sin R$, then $60 = \dfrac{1}{2} \cdot 10 \cdot 20 \sin R = 100 \sin R$. Thus, $\sin R = \dfrac{60}{100} = 0.6$. $\angle R = \sin^{-1} 0.6 \approx 37°$. Because $\angle R$ is obtuse, find the supplementary angle: $180° - 37° = 143°$.

Answer: $\angle R$ is 143°.

Problem 2: Find sin120°, sin125°, and sin145°.

Use supplementary angles: sin120° = sin(180° − 120°) = sin60° = $\frac{\sqrt{3}}{2}$; sin125° = sin(180° − 125°) = sin55° = 0.8192; sin145° = sin(180° − 145°) = sin35° = 0.5736.

Answer: sin120° = $\frac{\sqrt{3}}{2}$, sin125° = 0.8192, and sin145° = 0.5736

Problem 3: In ΔABC, ∠A is 45°, b is 4, and a = 2√2 . Find the number of solutions for the triangle.

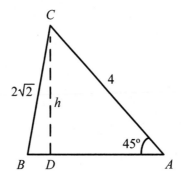

From ΔADC: $\frac{h}{4}$ = sin 45°, thus $h = 4 \cdot \sin 45° = 4 \cdot \frac{\sqrt{2}}{2} = 2\sqrt{2}.$ Since the altitude is equal to one of the sides, then the triangle is a right triangle.

Answer: There is one solution with a right triangle.

Problem 4: In a ΔABC, b is 7, ∠A is 40°, and ∠C is 28°. Find the length of a.

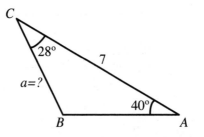

Find the third angle: 180° − (40° + 28°) = 180° − 68° = 112°. Use the law of sines:
$\frac{\sin 112°}{7} = \frac{\sin 40°}{a}$ or $a = \frac{7 \cdot \sin 40°}{\sin 112°}$. Then use supplementary angle

$\frac{7 \cdot \sin 40°}{\sin 68°} = \frac{7 \cdot 0.6428}{0.9272} \approx 4.9.$

Answer: The length of the side a is approximately 4.9.

Problem 5: Use the law of sines to show that there is no $\triangle PQS$ with $\angle P$ equaling 30°, p equaling 3, and q equaling 8.

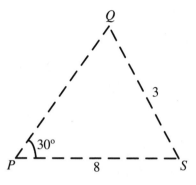

According to the law of sines: $\dfrac{\sin 30°}{3} = \dfrac{\sin Q}{8}$. Thus, $\sin Q = \dfrac{8 \sin 30°}{3} = \dfrac{8 \cdot \dfrac{1}{2}}{3} = \dfrac{4}{3} > 1.$

Since the sine of any angle cannot be greater than 1, $\triangle PQS$ does not exist.

Answer: $\triangle PQS$ does not exist.

Chapter 7: The Law of Cosines

Problem 1: In $\triangle ABC$, $\angle C$ measures 60°, a is 2, and b is 8. Find the length of side c.

Use the law of cosines for the side c: $c^2 = a^2 + b^2 - 2ab \cos C = 2^2 + 8^2 - 2 \cdot 2 \cdot 8 \cos 60°$

$= 4 + 64 - 32 \cdot \dfrac{1}{2} = 68 - 16 = 52$. Then $c = \sqrt{52} \approx 7.2$.

Answer: The length of side c is 7.2.

Problem 2: A triangle has sides of length 10, 16, and 20. Find the measure of its largest angle.

Since the largest angle in a triangle is always opposite to the largest side, you are looking for the angle that is opposite to the side with length 20, using

$\cos(\text{angle}) = \dfrac{(\text{adj.})^2 + (\text{adj.})^2 - (\text{opp.})^2}{2 \cdot (\text{adj.}) \cdot (\text{adj.})}$. Thus $\cos(\text{angle}) = \dfrac{(10)^2 + (16)^2 - (20)^2}{2 \cdot (10) \cdot (16)}$

$= \dfrac{100 + 256 - 400}{320} = \dfrac{-44}{320} = -0.1375.$

Then $\cos^{-1}(-0.1375) = 97.9°$.

Answer: The largest angle is equal to 97.9°.

Problem 3: After leaving the airport, a plane flies for 1 hour at a speed of 250 kilometers per hour on a course of 200°. Then it flies for 3 hours on a course of 340° at a speed of 200 kilometers per hour. At this time, how far from the airport is the plane?

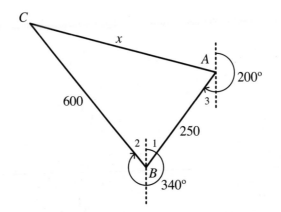

Designate point A as the airport, point B as where the plane changed heading, and point C as the plane's current location.

Distance $BC = 200$ km/h $\cdot$ $3h = 600$ km. In $\triangle ABC$, $\angle B = \angle 1 + \angle 2$. Two alternate interior angles $\angle 1 = \angle 3 = 200° - 180° = 20°$ and $\angle 2 = 360° - 340° = 20°$. Thus, $\angle B = \angle 1 + \angle 2 = 20° + 20° = 40°$. Use the law of cosines to find x:

$$x^2 = 250^2 + 600^2 - 2 \cdot 250 \cdot 600 \cdot \cos 40° = 62,500 + 360,000 - 300,000 \cdot 0.7660 = 192,700$$

$$x = \sqrt{192,700} = 439 \text{ km}$$

Answer: The plane is 439 kilometers away from the airport.

Problem 4: Sketch the following plot and find its area: from the southeast corner of the local park on Cherry Hill Road, proceed S78°W for 250 meters along the southern boundary of the park to the old oak tree. From the tree, proceed S15°E for 180 meters to Mulberry Lane, then N78°E along Mulberry Lane until it intersects Cherry Hill Road, and finally N30°E along Cherry Hill Road back to the starting point.

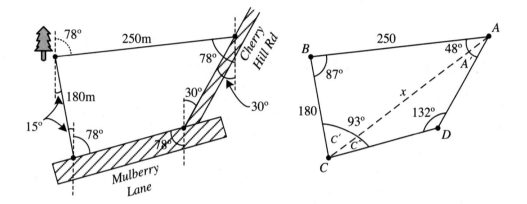

Using the law of cosines, find x: $x = \sqrt{250^2 + 180^2 - 2 \cdot 250 \cdot 180 \cos 87°} \approx 300$. Using the law of sines, find $\angle C'$: $\dfrac{\sin C'}{250} = \dfrac{\sin 87°}{300}$, thus $\sin C' = 0.8331$, thus $\angle C' = 56.4°$.

$\angle C'' = 93° - 56.4° = 36.6°$. $\angle A' = 180° - (132° + 36.6°) = 11.4°$. Using the law of sines, find AC: $\dfrac{\sin C''}{AD} = \dfrac{\sin 132°}{300}$, thus $AC = \dfrac{300 \cdot \sin 36.6°}{\sin 132°} \approx 241$. Area of CBA:

$A_1 = \dfrac{1}{2} \cdot 250 \cdot 180 \cdot \sin 87° \approx 22,468$. Area of CAD: $A_2 = \dfrac{1}{2} \cdot 300 \cdot 241 \cdot \sin 11.4° \approx 7,145$. Area of the plot: $A_1 + A_2 = 22,468 + 7,145 = 29,613$.

Answer: The plot area is 29,613 square meters.

Problem 5: Find the area of a triangle with sides of 5, 6, and 7 using Hero's formula.

Find s:

$$s = \frac{a+b+c}{2} = \frac{5+6+7}{2} = 9$$

Find the area:

$$A = \sqrt{9(9-5)(9-6)(9-7)} = \sqrt{216} \approx 14.7$$

Answer: The area is 14.7 square units.

Chapter 8: Angles and Rotations

Problem 1: Convert each degree measure to radians: 180°, 240°, 135°, 225°, and −135°. Leave answers in terms of π.

$$180° = 180° \cdot \frac{\pi}{180} = \pi$$

$$240° = 240 \cdot \frac{\pi}{180} = \frac{\overset{4}{\cancel{240}} \cdot \pi}{\underset{3}{\cancel{180}}} = \frac{4\pi}{3}$$

$$135° = 135 \cdot \frac{\pi}{180} = \frac{\overset{3}{\cancel{135}} \cdot \pi}{\underset{4}{\cancel{180}}} = \frac{3\pi}{4}$$

$$225° = 225 \cdot \frac{\pi}{180} = \frac{\overset{5}{\cancel{225}} \cdot \pi}{\underset{4}{\cancel{180}}} = \frac{5\pi}{4}$$

$$-135° = -135 \cdot \frac{\pi}{180} = -\frac{\overset{3}{\cancel{135}} \cdot \pi}{\underset{4}{\cancel{180}}} = -\frac{3\pi}{4}$$

Answer: The angles are π, $\dfrac{4\pi}{3}$, $\dfrac{3\pi}{4}$, $\dfrac{5\pi}{4}$, and $-\dfrac{3\pi}{4}$.

Problem 2: Convert each radian measure to degrees: $\dfrac{3\pi}{4}$, $\dfrac{11\pi}{6}$, $-\dfrac{7\pi}{6}$, $\dfrac{2\pi}{3}$, and $\dfrac{5\pi}{6}$.

$$\frac{3\pi}{4} = \frac{3\pi}{4} \cdot \frac{180}{\pi} = \frac{3 \cdot \overset{45}{\cancel{180}}}{\underset{1}{\cancel{4}}} = 135°$$

$$\frac{11\pi}{6} = \frac{11\pi}{6} \cdot \frac{180}{\pi} = \frac{11 \cdot \overset{30}{\cancel{180}}}{\underset{1}{\cancel{6}}} = 330°$$

$$-\frac{7\pi}{6} = -\frac{7\pi}{6} \cdot \frac{180}{\pi} = -\frac{7 \cdot \overset{30}{\cancel{180}}}{\underset{1}{\cancel{6}}} = -210°$$

$$\frac{2\pi}{3} = \frac{2\pi}{3} \cdot \frac{180}{\pi} = \frac{2 \cdot \overset{60}{\cancel{180}}}{\underset{1}{\cancel{3}}} = 120°$$

$$\frac{5\pi}{6} = \frac{5\pi}{6} \cdot \frac{180}{\pi} = \frac{5 \cdot \overset{30}{\cancel{180}}}{\underset{1}{\cancel{6}}} = 150°$$

Answer: The measures are 135°, 330°, –210°, 120°, and 150°.

Problem 3: Find two angles, one positive and one negative, that are coterminal with each given angle: 400° and $\dfrac{\pi}{4}$.

400°:

$400° - 360° = 40°$

$400° - 360° \rightarrow 40° - 360° = -320°$

$\dfrac{\pi}{4}$:

$\dfrac{\pi}{4} + 2\pi = \dfrac{\pi}{4} + \dfrac{8\pi}{4} = \dfrac{9\pi}{4}$

$\dfrac{\pi}{4} - 2\pi = \dfrac{\pi}{4} - \dfrac{8\pi}{4} = -\dfrac{7\pi}{4}$

Answer: The angles are 40° and –320° for the first angle and $\dfrac{9\pi}{4}$ and $-\dfrac{7\pi}{4}$ for the second one.

Problem 4: A race car moves around a circular track. Find the speed of the car if the track has a radius of 250 meters and the car covered a quarter of the track in 5 seconds.

The angle is 360° ÷ 4 = 90° or $\dfrac{\pi}{4}$. Find the distance it covered in 5 seconds:

$s = r\theta = 250 \cdot \dfrac{\pi}{4} \approx 196$ meters or 0.196 kilometers in 5 seconds. Then, in 1 hour, the car will travel 720 times farther (1 hour is 3,600 seconds; 3,600 ÷ 5 seconds = 720 times). Thus, the speed is 0.196 · 720 ≈ 140 kilometers per hour.

Answer: The car's speed is approximately 140 kilometers per hour.

Problem 5: The diameter of the moon is approximately 3,500 kilometers. Its apparent size from Earth is 0.0087 radians. How far is the moon from Earth?

You need to find the radius of the circle when you know the central angle (0.0087 radians) and the arc's length. Use the formula: $s = r \cdot \theta$, $r = \dfrac{s}{\theta} = \dfrac{3500}{0.0087} \approx 402,000$ km.

Answer: The distance to the moon is 402,000 kilometers.

Chapter 9: The Unit Circle Approach

Problem 1: Show that coordinates for the angle of $\dfrac{5\pi}{4}$ satisfy the unit circle's equation.

Coordinates are $\left(-\dfrac{\sqrt{2}}{2}, -\dfrac{\sqrt{2}}{2}\right)$, so $x^2 + y^2 = \left(-\dfrac{\sqrt{2}}{2}\right)^2 + \left(-\dfrac{\sqrt{2}}{2}\right)^2 = \dfrac{2}{4} + \dfrac{2}{4} = 1.$

Answer: The coordinates satisfy the unit circle equation.

Problem 2: Find the cotangent of $1,845°$.

Find the coterminal angle: $1,845° \div 360 = 5.125$. Then, $360 \cdot 5 = 1,800$ and $1,845 - 1,800 = 45$. Thus, the cotangent of $1,845°$ is the same as cotangent of $45°$:

$\cot 45° = \dfrac{\cos 45°}{\sin 45°} = \dfrac{\frac{\sqrt{2}}{2}}{\frac{\sqrt{2}}{2}} = \dfrac{\sqrt{2} \cdot 2}{\sqrt{2} \cdot 2} = 1$. Thus, $\cot 1,845° = 1$.

Answer: The cotangent of $1,845°$ is 1.

Problem 3: Determine whether the cosine of $25,000°$ is positive or negative.

Find the coterminal angle: $25,000 \div 360 = 69.444....$ Then, $360 \cdot 69 = 24,840$ and $25,000 - 24,840 = 160$. An angle of $160°$ is in the second quadrant, where the cosine is negative; therefore, the cosine of $25,000°$ is negative as well.

Answer: The cosine of $25,000°$ is negative.

Problem 4: Find the six trigonometric functions for $\dfrac{4\pi}{3}$.

Angle $\dfrac{4\pi}{3}$ is a multiple of angle $\dfrac{\pi}{3}$, which serves as a reference angle for $\dfrac{4\pi}{3}$ and is in quadrant III. Then, $\sin \dfrac{4\pi}{3} = -\dfrac{\sqrt{3}}{2}$, $\cos \dfrac{4\pi}{3} = -\dfrac{1}{2}$, $\tan \dfrac{4\pi}{3} = \sqrt{3}$, $\cot \dfrac{4\pi}{3} = \dfrac{\sqrt{3}}{3}$, $\csc \dfrac{4\pi}{3} = -\dfrac{2\sqrt{3}}{3}$, and $\sec \dfrac{4\pi}{3} = -2$.

Answer: $\sin \dfrac{4\pi}{3} = -\dfrac{\sqrt{3}}{2}$, $\cos \dfrac{4\pi}{3} = -\dfrac{1}{2}$, $\tan \dfrac{4\pi}{3} = \sqrt{3}$, $\cot \dfrac{4\pi}{3} = \dfrac{\sqrt{3}}{3}$, $\csc \dfrac{4\pi}{3} = -\dfrac{2\sqrt{3}}{3}$, and $\sec \dfrac{4\pi}{3} = -2$

Problem 5: Find the six trigonometric functions for $\dfrac{5\pi}{2}$.

$\dfrac{5\pi}{2} = \dfrac{4\pi}{2} + \dfrac{\pi}{2} = 2\pi + \dfrac{\pi}{2}$, so it's coterminal with $\dfrac{\pi}{2}$ and has the same values of its trigonometric functions:

$$\sin\dfrac{5\pi}{2} = 1, \ \cos\dfrac{5\pi}{2} = 0, \ \tan\dfrac{5\pi}{2} = \dfrac{\sin\dfrac{5\pi}{2}}{\cos\dfrac{5\pi}{2}} = \dfrac{1}{0} = \text{undefined},$$

$$\csc\dfrac{5\pi}{2} = 1, \ \sec\dfrac{5\pi}{2} = \dfrac{1}{\cos\dfrac{5\pi}{2}} = \dfrac{1}{0} = \text{undefined, and } \cot\dfrac{5\pi}{2} = \dfrac{\cos\dfrac{5\pi}{2}}{\sin\dfrac{5\pi}{2}} = \dfrac{0}{1} = 0$$

Answer: $\sin\dfrac{5\pi}{2} = 1, \ \cos\dfrac{5\pi}{2} = 0, \ \tan\dfrac{5\pi}{2} = \text{undefined}, \ \csc\dfrac{5\pi}{2} = 1, \ \sec\dfrac{5\pi}{2} = \text{undefined},$

and $\cot\dfrac{5\pi}{2} = 0$

Chapter 10: Trigonometric Functions of Any Angle

Problem 1: Find the sine, cosine, and tangent for an angle of 834°.

Find the coterminal angle: $834° - 2 \cdot 360° = 834° - 720° = 114°$. This is in the second quadrant. Find the reference angle: $180° - 114° = 66°$. Observing the signs for the trig functions in the second quadrant and taking the values of trig functions for the reference angle of 66°, you have: $\sin 834° = 0.9135$, $\cos 834° = -0.4067$, and $\tan 834° = -2.2460$.

Answer: $\sin 834° = 0.9135$, $\cos 834° = -0.4067$, and $\tan 834° = -2.2460$

Problem 2: Find the sine, cosine, and tangent for an angle of $\dfrac{34\pi}{3}$.

There were 5 full revolutions, since $34\pi \div 3 > 10\pi$, but less than 12. Only 10 (not 11) is evenly divisible by 2. Rewrite the angle as: $\dfrac{34\pi}{3} = \dfrac{30\pi}{3} + \dfrac{4\pi}{3} = 10\pi + \dfrac{4\pi}{3} = 5(2\pi) + \dfrac{4\pi}{3}$; therefore, the coterminal angle is $\dfrac{4\pi}{3}$, which is in the third quadrant. The reference angle is $\dfrac{4\pi}{3} - \pi = \dfrac{4\pi}{3} - \dfrac{3\pi}{3} = \dfrac{\pi}{3}$. Thus, observing the signs for the third quadrant and taking the values for the reference angle, you have $\sin\dfrac{34\pi}{3} = -\dfrac{\sqrt{3}}{2}$, $\cos\dfrac{34\pi}{3} = -\dfrac{1}{2}$, and $\tan\dfrac{34\pi}{3} = \sqrt{3}$.

Answer: $\sin\dfrac{34\pi}{3} = -\dfrac{\sqrt{3}}{2}$, $\cos\dfrac{34\pi}{3} = -\dfrac{1}{2}$, and $\tan\dfrac{34\pi}{3} = \sqrt{3}$

Problem 3: Find the sine, cosine, and tangent for an angle of 150°.

Find the reference angle: 180° – 150° = 30°. Using the correct signs for the second quadrant and the trig functions values for a special angle, you have $\sin 150° = \frac{1}{2}$, $\cos 150° = -\frac{\sqrt{3}}{2}$, and $\tan 150° = -\frac{\sqrt{3}}{3}$.

Answer: $\sin 150° = \frac{1}{2}$, $\cos 150° = -\frac{\sqrt{3}}{2}$, and $\tan 150° = -\frac{\sqrt{3}}{3}$

Problem 4: Find the cosecant of –132°.

Find the coterminal angle: –132° + 360° = 228°. 228° – 180° = 48°. sin48° = 0.7431, then sin228° = –0.7431. Thus, $\csc(-132°) = \frac{1}{\sin 228°} = -\frac{1}{0.7431} = -1.3456$.

Answer: csc(–132°) = –1.3456

Problem 5: Find the cotangent of –405°.

Find the coterminal angle: –405° + 360° = –45° + 360° = 315°. Find the reference angle: 360° – 315° = 45°. tan45° = 1, then tan(315°) = –1; therefore, tan(–405) = –1. Thus, $\cot(-405°) = \frac{1}{\tan(-405°)} = \frac{1}{-1} = -1$.

Answer: cot(–405 °) = –1

Chapter 11: Graphs of Sine and Cosine Functions

Problem 1: Sketch the graph of $y = 2\sin x$.

Answer: The amplitude is 2, so the graph will be stretched vertically and each y-coordinate of key points of the basic sine curve will be doubled:

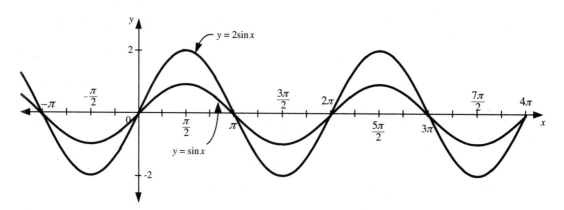

Problem 2: Sketch the graph of $y = \dfrac{1}{2}\cos x$.

Answer: The amplitude is $\dfrac{1}{2}$, so the graph will be shrunk vertically and each y-coordinate of key points of the basic cosine curve will be divided by 2.

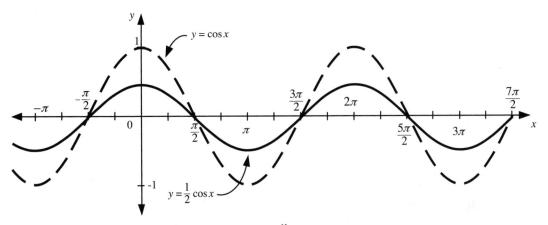

Problem 3: Sketch the graph of $y = 2\sin \dfrac{x}{4}$.

Answer: The amplitude is 2, so the graph will be stretched vertically and each y-coordinate of key points of the basic sine curve will be doubled. The period will be $2\pi \div \dfrac{1}{4} = 8\pi$, so the basic sine curve will be stretched out horizontally.

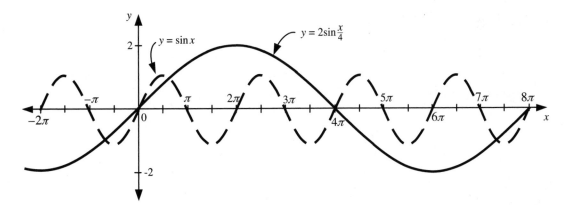

Problem 4: Sketch the graph of $y = 3\cos(x + \pi)$.

Answer: The amplitude is 3, so the graph will be stretched vertically and each y-coordinate of key points of the basic cosine curve will be tripled. The curve is also shifted to the left since $c = +\pi$. The first maximum is at $-\pi$ on the sketch (not at 0 as with the basic cosine curve).

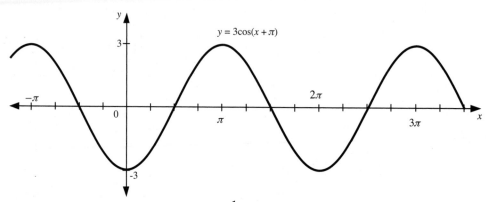

Problem 5: Sketch the graph of $y = -\dfrac{1}{2}\sin\dfrac{x}{2} + 3$.

Answer: The amplitude is $-\dfrac{1}{2}$, so the basic curve will be shrunk vertically and placed upside down due to the minus sign of the amplitude. The period will be $2\pi \div \dfrac{1}{2} = 4\pi$, so the curve will be stretched horizontally. Also, the curve will be shifted vertically and will go 3 units upward because $d = 3$.

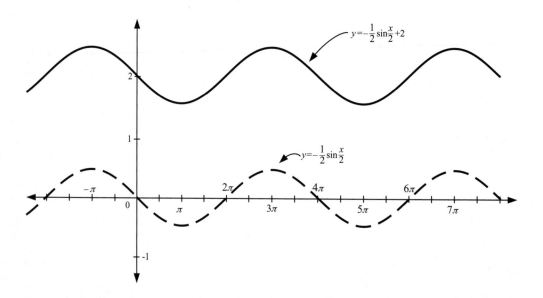

Chapter 12: Graphs of Other Trigonometric Functions

Problem 1: Sketch the graph of $2\tan\dfrac{x}{3}$.

Because the angle is multiplied by a fraction $\dfrac{1}{3}$, the period of the function is 3π (3 times more than the period of the basic tangent function). The first asymptote of the main cycle is $-\dfrac{\pi}{2}\cdot 3 = -\dfrac{3\pi}{2}$, and the second asymptote is $\dfrac{\pi}{2}\cdot 3 = \dfrac{3\pi}{2}$. The next asymptote can be found as $\dfrac{3\pi}{2} + 3\pi = \dfrac{3\pi}{2} + \dfrac{6\pi}{2} = \dfrac{9\pi}{2}$, and so forth. Because the whole function is multiplied by 2, the graph increases its steepness.

Answer: The graph is stretched along x-axis and more steep.

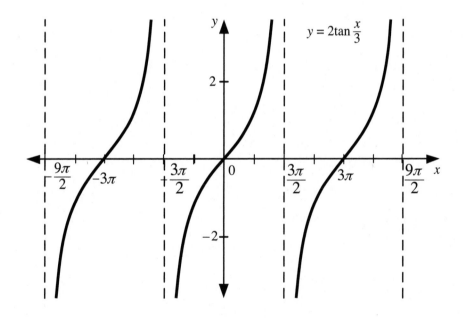

Problem 2: Sketch the graph of $\frac{1}{3}\cot x + 2$.

The period is the same: π. The graph is flatter since the cotangent is multiplied by a fraction. The graph is shifted upward 3 units and the point of inflection is at 3.

Answer: The graph is flatter and is shifted up 3 units along the y-axis.

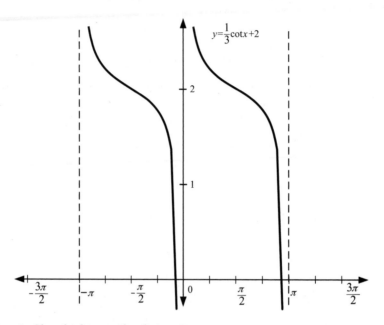

$y = \frac{1}{3}\cot x + 2$

Problem 3: Sketch the graph of $-3\tan 2x$.

The period is 2 times shorter than the one of the basic tangent graph and is equal to $\frac{\pi}{2}$. The first asymptote of the main cycle is $-\frac{\pi}{2} \div 2 = -\frac{\pi}{2} \cdot \frac{1}{2} = -\frac{\pi}{4}$ and the second asymptote is $\frac{\pi}{2} \div 2 = \frac{\pi}{2} \cdot \frac{1}{2} = \frac{\pi}{4}$. To find the next asymptote, add the new period: $\frac{\pi}{4} + \frac{\pi}{2} = \frac{\pi}{4} + \frac{2\pi}{4} = \frac{3\pi}{4}$, and so on. The graph is steeper since the function is multiplied by a whole number 3. Finally, the graph is flipped over the x-axis because of the minus sign in front of the function.

Answer: The graph has a shorter period, is steeper, and is flipped over the x-axis.

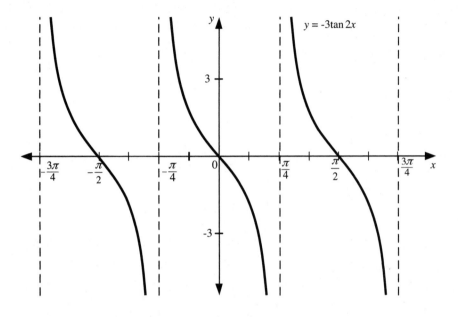

Problem 4: Sketch the graph of $3\csc(x + \pi)$.

The graph is shifted by π radians along the x-axis. The first asymptote is $0 - \pi = -\pi$ and the second asymptote is $\pi - \pi = 0$. Then, asymptotes repeat every π radians. Since the whole function is multiplied by 3, the distance between tops and bottoms is increased.

Answer: The graph is shifted to the left by π radians and the distance between the tops and bottoms is increased.

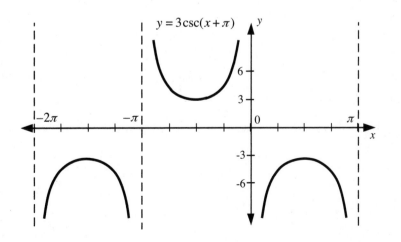

Problem 5: Sketch the graph of $\frac{1}{2}\sec 2x$.

The period is two times shorter and is equal to π. The first asymptote is $-\frac{\pi}{2} \div 2 = -\frac{\pi}{4}$ and the second asymptote is $\frac{\pi}{2} \div 2 = \frac{\pi}{4}$, and so forth. Because the whole function is multiplied by a fraction, the distance between the tops and bottoms is decreased.

Answer: Period is shorter and distance between tops and bottoms is decreased.

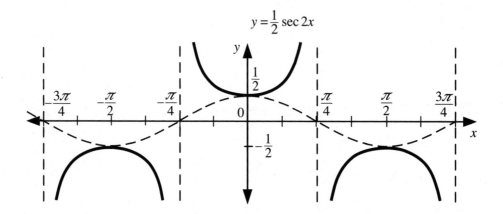

$$y = \frac{1}{2}\sec 2x$$

Chapter 13: Graphs of Inverse Trigonometric Functions

Problem 1: Find the exact value of $\arcsin\frac{1}{2}$.

By definition, $y = \arcsin\frac{1}{2}$ implies that $\sin y = \frac{1}{2}$. On the interval $-\frac{\pi}{2} \le y \le \frac{\pi}{2}$, only $\frac{\pi}{6}$ has sine that is equal to $\frac{1}{2}$. Thus, you can conclude that $\arcsin\frac{1}{2} = \frac{\pi}{6}$.

Answer: $\arcsin\frac{1}{2} = \frac{\pi}{6}$

Problem 2: Find the exact value of $\arctan\frac{\sqrt{3}}{3}$.

Because $\tan\left(\frac{\pi}{6}\right) = \frac{\sqrt{3}}{3}$ and $\frac{\pi}{6}$ lies in the range of the inverse tangent $\left[-\frac{\pi}{2}, \frac{\pi}{2}\right]$, you can conclude that $\arctan\frac{\sqrt{3}}{3} = \frac{\pi}{6}$.

Answer: $\arctan\frac{\sqrt{3}}{3} = \frac{\pi}{6}$

Problem 3: Find the exact value of cos(arccos(–0.1)).

Because –0.1 is in the domain of arccos$x$, you can apply the inverse property: cos(arccos(–0.1)) = –0.1.

Answer: –0.1

Problem 4: Find the exact value of $\arcsin(\sin 3\pi)$.

Even though 3π does not lie in the range $\left[-\dfrac{\pi}{2}, \dfrac{\pi}{2}\right]$, you can write that $(\sin 3\pi) = \sin 0$; therefore, $\arcsin(\sin 3\pi) = \arcsin(\sin 0) = 0$.

Answer: $\arcsin(\sin 3\pi) = 0$

Problem 5: Find the exact value of $\sec\left[\arctan\left(-\dfrac{3}{5}\right)\right]$.

Hint: make a sketch of a right triangle in the fourth quadrant. Denote an angle with a tangent of $\dfrac{-3}{5}$ as v, then you can state that $\angle v = \arctan\left(-\dfrac{3}{5}\right)$; therefore, $\tan v = -\dfrac{3}{5}$.

Since $\tan v$ is negative, $\angle v$ is a fourth-quadrant angle. Since $\tan = \dfrac{\text{opp.}}{\text{adj.}} = -\dfrac{3}{5}$, you can use the Pythagorean theorem to find the hypotenuse: $\text{hyp.} = \sqrt{(-3)^2 + 5^2} = \sqrt{34}$.

Then, $\sec v = \dfrac{1}{\cos v} = \dfrac{1}{\dfrac{\text{adj.}}{\text{hyp.}}} = \dfrac{\text{hyp.}}{\text{adj.}} = \dfrac{\sqrt{34}}{5}$.

Answer: $\sec\left[\arctan\left(-\dfrac{3}{5}\right)\right] = \dfrac{\sqrt{34}}{5}$

Chapter 14: Kaleidoscope of Identities

Problem 1: Use the fundamental identities to evaluate the other trigonometric functions if $\tan\theta = 2$ and $\sin\theta < 0$.

$$\cot\theta = \frac{1}{\tan\theta} = \frac{1}{2}$$

Use Pythagorean identity to find the secant: $\sec^2\theta = 1 + \tan^2\theta = 1 + 2^2 = 5$.

Because the tangent is positive and the sine is negative, this is a third-quadrant angle, where the secant is negative; thus, $\sec\theta = -\sqrt{5}$. Then, $\cos\theta = \frac{1}{\sec\theta} = \frac{1}{-\sqrt{5}} = -\frac{\sqrt{5}}{5}$.

Find the sine: $\tan\theta = \frac{\sin\theta}{\cos\theta}$, $\sin\theta = \tan\theta \cdot \cos\theta = 2 \cdot \left(-\frac{\sqrt{5}}{5}\right) = -\frac{2\sqrt{5}}{5}$. Lastly,

$$\csc\theta = \frac{1}{\sin\theta} = \frac{1}{-\frac{2\sqrt{5}}{5}} = -\frac{5}{2\sqrt{5}} = -\frac{\sqrt{5}}{2}.$$

Answer: $\sin\theta = -\frac{2\sqrt{5}}{5}$, $\cos\theta = -\frac{\sqrt{5}}{5}$, $\csc\theta = -\frac{\sqrt{5}}{2}$, $\sec\theta = -\sqrt{5}$, and $\cot\theta = \frac{1}{2}$

Problem 2: Factor the expression and use fundamental identities to simplify: $\cos^2\theta\csc^2\theta - \cos^2\theta$.

$$\cos^2\theta\csc^2\theta - \cos^2\theta = \cos^2\theta\left(\csc^2\theta - 1\right) = \cos^2\theta \cdot \cot^2\theta$$

Answer: $\cos^2\theta\csc^2\theta - \cos^2\theta = \cos^2\theta \cdot \cot^2\theta$

Problem 3: Factor the expression and use fundamental identities to simplify: $\cos^4\theta - \sin^4\theta$.

Use the difference of two squares formula and the Pythagorean identity:

$$\cos^4\theta - \sin^4\theta = \left(\sin^2\theta + \cos^2\theta\right)\left(\cos^2\theta - \sin^2\theta\right) = \cos^2\theta - \sin^2\theta$$

Answer: $\cos^4\theta - \sin^4\theta = \cos^2\theta - \sin^2\theta$

Problem 4: Subtract and simplify: $\cot\theta - \frac{\csc^2\theta}{\cot\theta}$.

$$\cot\theta - \frac{\csc^2\theta}{\cot\theta} = \frac{\cot\theta}{1} - \frac{\csc^2\theta}{\cot\theta} = \frac{\cot^2\theta - \csc^2\theta}{\cot\theta} = \frac{\cot^2\theta - \left(1 + \cot^2\theta\right)}{\cot\theta} = -\frac{1}{\cot\theta}$$

Answer: $\cot\theta - \frac{\csc^2\theta}{\cot\theta} = -\frac{1}{\cot\theta}$

Problem 5: Rewrite the expression so that it is not in fractional form: $\dfrac{7}{\sec\theta - \tan\theta}$.

Use reciprocal and quotient identities and combine fractions in the denominator:

$$\frac{7}{\sec\theta - \tan\theta} = \frac{7}{\dfrac{1}{\cos\theta} - \dfrac{\sin\theta}{\cos\theta}} = \frac{7}{\dfrac{1-\sin\theta}{\cos\theta}} = \frac{7\cos\theta}{1-\sin\theta}$$

Multiply the numerator and the denominator by the conjugate and use the difference of two squares formula in the denominator:

$$\frac{7\cos\theta}{1-\sin\theta} = \frac{7\cos\theta(1+\sin\theta)}{(1-\sin\theta)(1+\sin\theta)} = \frac{7\cos\theta(1+\sin\theta)}{1-\sin^2\theta}$$

Use the Pythagorean identity in the denominator and cancel common factors:

$$\frac{7\cos\theta(1+\sin\theta)}{1-\sin^2\theta} = \frac{7\cos\theta(1+\sin\theta)}{\cos^2\theta} = \frac{7(1+\sin\theta)}{\cos\theta}$$

Finally, separate fractions and use reciprocal and quotient identities:

$$\frac{7(1+\sin\theta)}{\cos\theta} = 7\left(\frac{1}{\cos\theta} + \frac{\sin\theta}{\cos\theta}\right) = 7\sec\theta + \tan\theta$$

Answer: $\dfrac{7}{\sec\theta - \tan\theta} = 7\sec\theta + \tan\theta$

Chapter 15: Verifying Trigonometric Identities

Problem 1: Verify the identity $\dfrac{\sin\theta\tan\theta}{1-\cos\theta} - 1 = \sec\theta$.

Multiply a fraction by a conjugate and use a difference of two squares formula:

$$\frac{\sin\theta\tan\theta}{1-\cos\theta} - 1 = \frac{\sin\theta\tan\theta(1+\cos\theta)}{(1-\cos\theta)(1+\cos\theta)} - 1 = \frac{\sin\theta\tan\theta(1+\cos\theta)}{1-\cos^2\theta} - 1$$

Use the Pythagorean identity for the denominator, use the tangent ratio in the numerator, and rearrange the complex fraction:

$$\frac{\sin\theta\tan\theta(1+\cos\theta)}{1-\cos^2\theta} - 1 = \frac{\sin\theta\cdot\dfrac{\sin\theta}{\cos\theta}(1+\cos\theta)}{\sin^2\theta} - 1 = \frac{\sin^2\theta(1+\cos\theta)}{\cos\theta\sin^2\theta} - 1$$

Reduce and subtract two fractions:

$$\frac{\sin^2\theta(1+\cos\theta)}{\cos\theta\sin^2\theta} - 1 = \frac{1+\cos\theta}{\cos\theta} - 1 = \frac{1+\cos\theta-\cos\theta}{\cos\theta} = \frac{1}{\cos\theta} = \sec\theta$$

Answer: $\dfrac{\sin\theta\tan\theta}{1-\cos\theta} - 1 = \sec\theta$

Problem 2: Verify the identity $\dfrac{\csc\theta - 1}{1 - \sin\theta} = \csc\theta$.

Work with the left side, use reciprocal identity, subtract fractions in the numerator, and rearrange the complex fraction:

$$\frac{\csc\theta - 1}{1 - \sin\theta} = \frac{\dfrac{1}{\sin\theta} - 1}{1 - \sin\theta} = \frac{\dfrac{1 - \sin\theta}{\sin\theta}}{1 - \sin\theta} = \frac{1 - \sin\theta}{\sin\theta(1 - \sin\theta)}$$

Reduce and use reciprocal identity:

$$\frac{1 - \sin\theta}{\sin\theta(1 - \sin\theta)} = \frac{1}{\sin\theta} = \csc\theta$$

Answer: You verified $\dfrac{\csc\theta - 1}{1 - \sin\theta} = \csc\theta$.

Problem 3: Verify the identity $\dfrac{1 + \sec(-\theta)}{\sin(-\theta) + \tan(-\theta)} = -\csc\theta$.

Work with the left side and use negative angle identities, reciprocal identities, and quotient identities to convert everything to sines and cosines:

$$\frac{1 + \sec(-\theta)}{\sin(-\theta) + \tan(-\theta)} = \frac{1 + \sec\theta}{-\sin\theta - \tan\theta} = \frac{1 + \dfrac{1}{\cos\theta}}{-\left(\sin\theta + \dfrac{\sin\theta}{\cos\theta}\right)}$$

Add fractions both in the numerator and denominator and rearrange the complex fraction. Then factor and reduce:

$$\frac{1 + \dfrac{1}{\cos\theta}}{-\left(\sin\theta + \dfrac{\sin\theta}{\cos\theta}\right)} = \frac{\dfrac{\cos\theta + 1}{\cos\theta}}{-\left(\dfrac{\sin\theta\cos\theta + \sin\theta}{\cos\theta}\right)} = \frac{\cos\theta + 1}{-\sin\theta(\cos\theta + 1)} = -\frac{1}{\sin\theta} = -\csc\theta$$

Answer: You verified that $\dfrac{1 + \sec(-\theta)}{\sin(-\theta) + \tan(-\theta)} = -\csc\theta$.

Problem 4: Verify the identity $\sqrt{\dfrac{1 + \cos\theta}{1 - \cos\theta}} = \dfrac{1 + \cos\theta}{|\sin\theta|}$.

Square both sides and then work with each side separately.

The left side:

$$\left(\sqrt{\frac{1 + \cos\theta}{1 - \cos\theta}}\right)^2 = \frac{(1 + \cos\theta)(1 + \cos\theta)}{(1 - \cos\theta)(1 + \cos\theta)} = \frac{(1 + \cos\theta)^2}{1 - \cos^2\theta} = \frac{1 + 2\cos\theta + \cos^2\theta}{\sin^2\theta}$$

The right side:

$$\left(\frac{1+\cos\theta}{|\sin\theta|}\right)^2 = \frac{1+2\cos\theta+\cos^2\theta}{\sin^2\theta}$$

Because both sides are equal to the same expression, you proved the identity.

Answer: You verified that $\sqrt{\dfrac{1+\cos\theta}{1-\cos\theta}} = \dfrac{1+\cos\theta}{|\sin\theta|}$.

Problem 5: Verify the identity $\dfrac{\cot\theta}{\csc\theta-1} = \dfrac{\csc\theta+1}{\cot\theta}$.

Work with both sides. On the left side, convert everything to the sines and cosines, combine terms in the denominator and rearrange a complex fraction:

$$\frac{\cot\theta}{\csc\theta-1} = \frac{\dfrac{\cos\theta}{\sin\theta}}{\dfrac{1}{\sin\theta}-1} = \frac{\dfrac{\cos\theta}{\sin\theta}}{\dfrac{1-\sin\theta}{\sin\theta}} = \frac{\cos\theta}{1-\sin\theta}$$

Multiply by the conjugate, use the difference of two squares, apply the Pythagorean identity, and reduce:

$$\frac{\cos\theta}{1-\sin\theta} = \frac{\cos\theta(1+\sin\theta)}{(1-\sin\theta)(1+\sin\theta)} = \frac{\cos\theta(1+\sin\theta)}{1-\sin^2\theta} = \frac{\cos\theta(1+\sin\theta)}{\cos^2\theta} = \frac{(1+\sin\theta)}{\cos\theta}$$

Work with the right side and convert to the sines and cosines, and work with a complex fraction:

$$\frac{\csc\theta+1}{\cot\theta} = \frac{\dfrac{1}{\sin\theta}+1}{\dfrac{\cos\theta}{\sin\theta}} = \frac{\dfrac{1+\sin\theta}{\sin\theta}}{\dfrac{\cos\theta}{\sin\theta}} = \frac{1+\sin\theta}{\cos\theta}$$

Because both sides are equal to the same expression, you proved the identity.

Answer: You verified the identity $\dfrac{\cot\theta}{\csc\theta-1} = \dfrac{\csc\theta+1}{\cot\theta}$.

Chapter 16: Solving Trigonometric Equations

Problem 1: Solve $3\sec^2 x - 4 = 0$.

Add 4 to both sides to get $3\sec^2 x = 4$, then divide by 3 to obtain $\sec^2 x = \dfrac{4}{3}$. Extract the square root to get $\sec x = \pm\sqrt{\dfrac{4}{3}} = \pm\dfrac{2}{\sqrt{3}} = \pm\dfrac{2\sqrt{3}}{3}$. Solutions for the interval $[0, \pi]$ are $\dfrac{\pi}{6}$ and $\dfrac{5\pi}{6}$. Because the period of the secant is π, general solutions are $\dfrac{\pi}{6} + n\pi$ and $\dfrac{5\pi}{6} + n\pi$. Note that when $n = 1$, the general solutions cover the angles with a secant of $-\dfrac{2\sqrt{3}}{3}$ $\left(\dfrac{\pi}{6} + \pi = \dfrac{\pi}{6} + \dfrac{6\pi}{6} = \dfrac{7\pi}{6}\right.$ and $\dfrac{5\pi}{6} + \pi = \dfrac{5\pi}{6} + \dfrac{6\pi}{6} = \dfrac{11\pi}{6}$, and their coterminals when n is more than 1).

Answer: $x = \dfrac{\pi}{6} + n\pi$ and $x = \dfrac{5\pi}{6} + n\pi$

Problem 2: Solve $\sin^2 x = 3\cos^2 x$.

Subtract $3\cos^2 x$ from both sides, use the Pythagorean identity, and collect like terms: $\sin^2 x - 3\cos^2 x = 0$, $\sin^2 x - 3(1 - \sin^2 x) = 0$, $4\sin^2 x - 3 = 0$, and $4\sin^2 x = 3$. Extract a square root: $\sin x = \pm\dfrac{\sqrt{3}}{2}$. Solutions for the interval $[0, \pi]$ are $\dfrac{\pi}{3}$ and $\dfrac{2\pi}{3}$. The period of the sine is 2π, but you add only $n\pi$ to get general solutions $\dfrac{\pi}{3} + n\pi$ and $\dfrac{2\pi}{3} + n\pi$, which cover the angles with a sine of $\pm\dfrac{2\sqrt{3}}{3}$ $\left(\dfrac{\pi}{3} + \pi = \dfrac{\pi}{3} + \dfrac{3\pi}{3} = \dfrac{4\pi}{3}\right.$ and $\dfrac{2\pi}{3} + \pi = \dfrac{2\pi}{3} + \dfrac{3\pi}{3} = \dfrac{5\pi}{3}$, and their coterminals when n is more than 1).

Answer: $x = \dfrac{\pi}{3} + n\pi$ and $x = \dfrac{5\pi}{3} + n\pi$

Problem 3: Solve $2\sin^2 x + 3\sin x + 1 = 0$ in the interval $[0, 2\pi]$.

Factor as a trinomial: $(2\sin x + 1)(\sin x + 1) = 0$. After setting each factor equal to zero, you get $\sin x = -\dfrac{1}{2}$ and $\sin x = -1$. In the interval $[0, 2\pi]$, there are three solutions: $\dfrac{7\pi}{6}$, $\dfrac{2\pi}{3}$, and $\dfrac{11\pi}{6}$.

Answer: $x = \dfrac{7\pi}{6}$, $x = \dfrac{2\pi}{3}$, and $x = \dfrac{11\pi}{6}$

Problem 4: Solve $\sin x = \cos x$. Hint: square both sides.

Square both sides, use the Pythagorean identity, and collect like terms: $\sin^2 x = \cos^2 x$, $\sin^2 x - \cos^2 x = \sin^2 x - (1 - \sin^2 x) = 0$, $2\sin^2 x - 1 = 0$, and $\sin^2 x = \dfrac{1}{2}$. Extract a square root: $\sin x = \pm\dfrac{\sqrt{2}}{2}$. Solutions for the interval $[0, \pi]$ are $\dfrac{\pi}{4}$ and $\dfrac{3\pi}{4}$. The period of the sine is 2π, but you add only π to get general solutions $\dfrac{\pi}{4}+n\pi$ and $\dfrac{3\pi}{4}+n\pi$, which cover the angles with a sine of $-\dfrac{\sqrt{2}}{2}$ ($\dfrac{\pi}{4}+\pi=\dfrac{\pi}{4}+\dfrac{4\pi}{4}=\dfrac{5\pi}{4}$, $\dfrac{3\pi}{4}+\pi=\dfrac{3\pi}{4}+\dfrac{4\pi}{4}=\dfrac{7\pi}{4}$, and their coterminals when n more than 1.)

Answer: $x=\dfrac{\pi}{4}+n\pi$ and $x=\dfrac{3\pi}{4}+n\pi$

Problem 5: Solve $3\tan^3 x - 3\tan^2 x - \tan x + 1 = 0$. Hint: use factoring by grouping.

Factor by grouping and set each factor equal to zero: $3\tan^2 x(\tan x - 1) - (\tan x - 1) = 0$, $(\tan x - 1)(3\tan^2 x - 1) = 0$.

First factor is $\tan x = 1$; its solution is $x = \dfrac{\pi}{4} + n\pi$. Second factor is $3\tan^2 x = 1$ or $\tan x = \pm\dfrac{\sqrt{3}}{3}$. Solutions for the interval $[0, \pi]$ are $\dfrac{\pi}{6}$ and $\dfrac{5\pi}{6}$. You add $n\pi$ to get general solutions $\dfrac{\pi}{6}+n\pi$ and $\dfrac{5\pi}{6}+n\pi$.

They cover the angles with tangent of $\pm\dfrac{\sqrt{3}}{3}$ ($\dfrac{\pi}{6}+\pi=\dfrac{\pi}{6}+\dfrac{6\pi}{6}=\dfrac{7\pi}{6}$ and $\dfrac{5\pi}{6}+\pi=\dfrac{5\pi}{6}+\dfrac{6\pi}{6}=\dfrac{11\pi}{6}$, and their coterminals when n is more than 1).

Answer: $x=\dfrac{\pi}{4}+n\pi$, $\dfrac{\pi}{6}+n\pi$, and $\dfrac{5\pi}{6}+n\pi$

Chapter 17: Sum and Difference Formulas

Problem 1: Find the cosine of $165°$.

Use the facts that $165° = 135° + 30°$ and that angle of $45°$ is the reference angle for $135°$. Thus $\sin 135° = \dfrac{\sqrt{2}}{2}$ and $\cos 135° = -\dfrac{\sqrt{2}}{2}$ since $135°$ is a second quadrant angle.

Use the sum formula: $\cos 165° = \cos(135° + 30°) = \cos 135° \cos 30° - \sin 135° \sin 30° =$

$$= -\frac{\sqrt{2}}{2}\cdot\frac{\sqrt{3}}{2} - \frac{\sqrt{2}}{2}\cdot\frac{1}{2} = -\frac{\sqrt{6}}{4} - \frac{\sqrt{2}}{4} = \frac{\sqrt{6}-\sqrt{2}}{4}.$$

Answer: $\cos 165° = \dfrac{\sqrt{6}-\sqrt{2}}{4}$

Problem 2: Verify the identity of $\sin\left(\dfrac{\pi}{2}+x\right) = \cos x$.

Use the sum formula for sine on the left:

$$\sin\left(\frac{\pi}{2}+x\right) = \sin\frac{\pi}{2}\cos x + \cos\frac{\pi}{2}\sin x = 1\cdot\cos x + 0\cdot\sin x = \cos x$$

Answer: You verified that $\sin\left(\dfrac{\pi}{2}+x\right) = \cos x$.

Problem 3: Find all solutions of the equation $\sin\left(x+\dfrac{\pi}{4}\right)+\sin\left(x-\dfrac{\pi}{4}\right)=-1$ in the interval $[0,2\pi)$.

On the left side, use the sum and difference formulas for sines, and then collect like terms: $\sin x \cos\dfrac{\pi}{4}+\cos x \sin\dfrac{\pi}{4}+\sin x \cos\dfrac{\pi}{4}-\cos x \sin\dfrac{\pi}{4}=-1,\ 2\sin x \cos\dfrac{\pi}{4}=-1,$

$2\sin x \cdot \dfrac{\sqrt{2}}{2}=-1,\ \sqrt{2}\sin x=-1.$ Then, $\sin x=-\dfrac{\sqrt{2}}{2}.$ In the interval $[0,2\pi)$, there two

solutions: $\dfrac{5\pi}{4}$ and $\dfrac{7\pi}{4}.$

Answer: $x=\dfrac{5\pi}{4}$ and $x=\dfrac{7\pi}{4}$

Problem 4: Find the tangent of $255°$.

$$255° = 300° - 45°$$

Apply the difference formula for tangent and use the tangent values for angle of $60°$, because this is a reference angle for $300°$.

$$\tan 255° = \tan\left(300°-45°\right)=\frac{\tan 300° - \tan 45°}{1+\tan 300° \tan 45°}=\frac{\tan 60° - \tan 45°}{1+\tan 60° \tan 45°}=\frac{\sqrt{3}-1}{1+\sqrt{3}}$$

Answer: $\tan 255° = \dfrac{\sqrt{3}-1}{1+\sqrt{3}}$

Problem 5: Find $\sin(v-u)$, if $\sin u = \dfrac{7}{25},\ \dfrac{\pi}{2}<u<\pi$ and $\cos v = \dfrac{4}{5},\ \dfrac{3\pi}{2}<v<2\pi.$

Using the Pythagorean identity find $\cos u$ and $\sin v$:

$$\cos u = \pm\sqrt{1-\left(\frac{7}{25}\right)^2}=\pm\sqrt{\frac{576}{625}}=\pm\frac{24}{25}$$

Since $\angle u$ is a second-quadrant angle, use a negative value:

$$\sin v = \pm\sqrt{1-\left(\frac{4}{5}\right)^2}=\pm\sqrt{\frac{9}{25}}=\pm\frac{3}{5}$$

Because $\angle v$ is a fourth-quadrant angle, use a negative value. Use the difference formula:

$$\sin\left(v-u\right)=-\frac{3}{5}\cdot\left(-\frac{24}{25}\right)-\left(\frac{4}{5}\right)\!\left(\frac{7}{25}\right)=\frac{72-28}{125}=\frac{44}{125}$$

Answer: $\sin\left(v-u\right)=\dfrac{44}{125}$

Chapter 18: Double-Angle and Power-Reducing Formulas

Problem 1: Find the exact value of the expression $\dfrac{2\sin\dfrac{\pi}{12}\cos\dfrac{\pi}{12}}{1-2\sin^2\dfrac{\pi}{12}}$.

Use double-angle formulas for the sine in the numerator and for the cosines in the denominator:

$$\frac{2\sin\dfrac{\pi}{12}\cos\dfrac{\pi}{12}}{1-2\sin^2\dfrac{\pi}{12}} = \frac{\sin\dfrac{\pi}{6}}{\cos\dfrac{\pi}{6}} = \tan\frac{\pi}{6} = \frac{\sqrt{3}}{3}.$$

Answer: The exact value is $\dfrac{\sqrt{3}}{3}$.

Problem 2: Solve the equation $\sin x \cos x = 0.25$ on the interval $[0, 2\pi]$.

Multiply each side by 2 to get the double-angle formula for the sine on the left side: $2 \cdot \sin x \cos x = 2 \cdot 0.25 = 0.5$. Use the formula: $\sin 2x = \dfrac{1}{2}$. On the interval $[0, 2\pi]$, there are two solutions: $2x = \dfrac{\pi}{6}$ and $2x = \dfrac{5\pi}{6}$. Then, $x = \dfrac{\pi}{12}$ and $x = \dfrac{5\pi}{12}$.

Answer: $x = \dfrac{\pi}{12}$ and $x = \dfrac{5\pi}{12}$

Problem 3: Find the exact value of the expression $\dfrac{2\tan 15°}{1-\tan^2 15°}$.

Use the double-angle formula for the tangent: $\dfrac{2\tan 15°}{1-\tan^2 15°} = \tan 30° = \dfrac{\sqrt{3}}{3}$.

Answer: The exact value is $\dfrac{\sqrt{3}}{3}$.

Problem 4: Find the exact values of $\sin 2u$, $\cos 2u$, and $\tan 2u$ if $\sin u = \dfrac{3}{5}$ and $0 < u < \dfrac{\pi}{2}$.

Find $\cos u$ first using the Pythagorean identity: $\cos u = \pm\sqrt{1-\left(\dfrac{3}{5}\right)^2} = \pm\sqrt{\dfrac{16}{25}} = \pm\dfrac{4}{5}$. Since the angle is a first-quadrant angle, use the positive value of the cosine. Use double-angle formulas to find the following trig functions:

$$\sin 2u = 2\sin u\cos u = 2 \cdot \frac{3}{5} \cdot \frac{4}{5} = \frac{24}{25}$$

$$\cos 2u = 1 - 2\sin^2 u = 1 - 2\left(\frac{3}{5}\right)^2 = 1 - 2 \cdot \frac{9}{25} = 1 - \frac{18}{25} = \frac{7}{25}$$

$$\tan 2u = \frac{\sin 2u}{\cos 2u} = \frac{\dfrac{24}{25}}{\dfrac{7}{25}} = \frac{24}{7}$$

Answer: $\sin 2u = \dfrac{24}{25}$, $\cos 2u = \dfrac{7}{25}$, and $\tan 2u = \dfrac{24}{7}$

Problem 5: Use the power-reducing formula to rewrite the expression $\sin^4 x$ in terms of the first power of the cosine.

Use power-reducing formula for the sine:

$$\sin^4 x = \left(\sin^2 x\right)^2 = \left(\frac{1 - \cos 2x}{2}\right)^2 = \frac{1 - 2\cos 2x + \cos^2 2x}{4}$$

Use power-reducing formula for the cosine:

$$\frac{1}{4}\left(1 - 2\cos 2x + \frac{1 + \cos 4x}{2}\right) = \frac{1}{4} - \frac{1}{2}\cos 2x + \frac{1}{8} + \frac{1}{8}\cos 4x$$

Collect like terms and factor:

$$\frac{3}{8} - \frac{1}{2}\cos 2x + \frac{1}{8}\cos 4x = \frac{1}{8}\left(3 - 4\cos 2x + \cos 4x\right)$$

Answer: $\sin^4 x = \dfrac{1}{8}\left(3 - 4\cos 2x + \cos 4x\right)$

Chapter 19: Half-Angle and Product-to-Sum Formulas

Problem 1: Find the value of $\sin\dfrac{x}{2}$ if $\cos x = \dfrac{12}{13}$ and $0 < x < \dfrac{\pi}{2}$.

Use half-angle formula and plug in the value for the cosine:

$$\sin\frac{x}{2} = \pm\sqrt{\frac{1 - \cos x}{2}} = \sqrt{\frac{1 - \dfrac{12}{13}}{2}} = \sqrt{\frac{1}{26}} \approx 0.1961$$

You used the positive sign since $\dfrac{x}{2}$ is a first-quadrant angle.

Answer: $\sin\dfrac{x}{2} \approx 0.1961$

Problem 2: Determine the exact value of the tangent of $\dfrac{\pi}{8}$.

Use half-angle formula, plug in the value for the tangent of the special angle, and simplify the complex fraction:

$$\tan\frac{\pi}{8} = \frac{1-\cos\dfrac{\pi}{4}}{\sin\dfrac{\pi}{4}} = \frac{1-\dfrac{\sqrt{2}}{2}}{\dfrac{\sqrt{2}}{2}} = \frac{\dfrac{2-\sqrt{2}}{2}}{\dfrac{\sqrt{2}}{2}} = \frac{2-\sqrt{2}}{\sqrt{2}}$$

Rationalize the denominator by multiplying by $\sqrt{2}$ and simplify:

$$\frac{\left(2-\sqrt{2}\right)\cdot\sqrt{2}}{\sqrt{2}\cdot\sqrt{2}} = \frac{2\sqrt{2}-2}{2} = \frac{2\left(\sqrt{2}-1\right)}{2} = \sqrt{2}-1$$

Answer: $\tan\dfrac{\pi}{8} = \sqrt{2}-1$

Problem 3: Use the product-to-sum formulas to write $6\sin\dfrac{\pi}{4}\cos\dfrac{\pi}{4}$ as a sum.

Use the formula and simplify:

$$6\sin\frac{\pi}{4}\cos\frac{\pi}{4} = 6\left[\frac{1}{2}\left(\sin\left(\frac{\pi}{4}+\frac{\pi}{4}\right)+\sin\left(\frac{\pi}{4}-\frac{\pi}{4}\right)\right)\right] = 6\left[\frac{1}{2}\left(\sin\frac{\pi}{2}+\sin 0\right)\right] = 3\left(\sin\frac{\pi}{2}+\sin 0\right)$$

Answer: $6\sin\dfrac{\pi}{4}\cos\dfrac{\pi}{4} = 3\left(\sin\dfrac{\pi}{2}+\sin 0\right)$

Problem 4: Verify the identity $\dfrac{\cos 4x - \cos 2x}{2\sin 3x} = -\sin x$.

Use sum-to-product formula and simplify the left side:

$$\frac{\cos 4x - \cos 2x}{2\sin 3x} = \frac{-2\sin\dfrac{4x+2x}{2}\cdot\sin\dfrac{4x-2x}{2}}{2\sin 3x} = \frac{-2\sin 3x\sin x}{2\sin 3x} = -\sin x$$

Answer: You verified that $\dfrac{\cos 4x - \cos 2x}{2\sin 3x} = -\sin x$.

Problem 5: Solve the equation $\sin 6x - \sin 4x = 0$ in the interval $[0, 2\pi)$.

Use sum-to-product formula on the left side and then set each factor equal to zeroto obtain solutions on the interval $[0, 2\pi)$: $\sin 6x - \sin 4x = 2\cos\dfrac{6x+4x}{2}\sin\dfrac{6x-4x}{2} = 0$,

$2\cos(5x)\sin(x) = 0$; thus $\cos 5x = 0$. Then $5x = \dfrac{\pi}{2}$ and $5x = \dfrac{3\pi}{2}$, or $x = \dfrac{\pi}{10}$ and $x = \dfrac{3\pi}{10}$. For $\sin x = 0$, $x = 0$ and $x = \pi$.

Answer: $x = 0$, $x = \dfrac{\pi}{10}$, $x = \dfrac{3\pi}{10}$, and $x = \pi$

Chapter 20: Polar Coordinates

Problem 1: Find three additional polar representations of the point $\left(2, \dfrac{3\pi}{4}\right)$, using $-2\pi < \theta < 2\pi$.

To find the three additional polar representations, we need to find an alternative polar representation in the form of (r, θ) and two more in the form of $(-r, \theta')$ with θ' being an angle "opposite" to the original one on the coordinate plane.

$$\frac{3\pi}{4} - 2\pi = -\frac{5\pi}{4} \rightarrow \left(2, -\frac{5\pi}{4}\right)$$

$$\frac{3\pi}{4} + \pi = \frac{7\pi}{4} \rightarrow \left(-2, \frac{7\pi}{4}\right)$$

$$\frac{3\pi}{4} - \pi = -\frac{\pi}{4} \rightarrow \left(-2, -\frac{\pi}{4}\right)$$

Answer: The additional polar representations are $\left(2, -\dfrac{5\pi}{4}\right)$, $\left(-2, \dfrac{7\pi}{4}\right)$, and $\left(-2, -\dfrac{\pi}{4}\right)$.

Problem 2: Find the corresponding rectangular coordinates for the point $\left(-1, \dfrac{5\pi}{4}\right)$.

To convert from rectangular coordinates to polar ones, let's use the formulas $x = r\cos\theta$ and $y = r\sin\theta$.

$$x = (-1)\cos\left(\frac{5\pi}{4}\right) = -\left(-\frac{1}{\sqrt{2}}\right) = \frac{1}{\sqrt{2}}$$

$$y = (-1)\sin\left(\frac{5\pi}{4}\right) = -\left(-\frac{1}{\sqrt{2}}\right) = \frac{1}{\sqrt{2}}$$

Answer: The corresponding rectangular coordinates for the point are $\left(\dfrac{1}{\sqrt{2}}, \dfrac{1}{\sqrt{2}}\right)$.

Problem 3: Find the corresponding polar coordinates for the point $\left(\dfrac{1}{2}, -\dfrac{\sqrt{3}}{2}\right)$.

In order to find the polar coordinates, let's use the formulas $r = \sqrt{x^2 + y^2}$ and $\theta = \tan^{-1}\left(\dfrac{y}{x}\right)$.

$$r = \sqrt{\left(\frac{1}{2}\right)^2 + \left(-\frac{\sqrt{3}}{2}\right)^2} = \sqrt{\frac{1}{4} + \frac{3}{4}} = \sqrt{\frac{4}{4}} = 1$$

$$\theta = \tan^{-1}\left(\frac{\left(-\frac{\sqrt{3}}{2}\right)}{\left(\frac{1}{2}\right)}\right) = \tan^{-1}\left(-\sqrt{3}\right) = -\frac{\pi}{3}$$

Answer: The corresponding polar coordinates for the point are $\left(1, -\frac{\pi}{3}\right)$.

Problem 4: Find the corresponding polar coordinates for the point (–3,4).

In order to find the polar coordinates, let's use the formulas $r = \sqrt{x^2 + y^2}$ and $\theta = \tan^{-1}\left(\frac{y}{x}\right)$.

$$r = \sqrt{3^2 + 4^2} = \sqrt{9 + 16} = \sqrt{25} = 5$$
$$\theta = \tan^{-1}\left(-\frac{4}{3}\right) = 126.87°$$

Note that we have chosen an angle that lies in the same quadrant as where the point (–3,4) is located.

Answer: The corresponding polar coordinates for the point are (5,126.87°).

Problem 5: Convert the polar equation $r = 1 - \sin\theta$ to rectangular form.

Let's utilize equations $y = r\sin\theta$ and $r = \sqrt{x^2 + y^2}$. You can convert the first equation into the form $\frac{y}{r} = \sin\theta$ and the second one into $r^2 = x^2 + y^2$.

Plugging $\frac{y}{r} = \sin\theta$ into the original problem yields $r = 1 - \frac{y}{r}$ after which we can multiply both sides by r to obtain $r^2 = r - y$. After plugging in the equations for both r^2 and r, you obtain $x^2 + y^2 = \sqrt{x^2 + y^2} - y$.

Answer: The rectangular form of the equation is $x^2 + y^2 = \sqrt{x^2 + y^2} - y$.

Chapter 21: Complex Numbers and Operations with Them

Problem 1: Express $4cis45°$ in rectangular form.

Recall that $rcis\theta = r(\cos\theta + i\sin\theta)$. In the case of this problem,

$$4cis45° = 4(\cos 45° + i\sin 45°) = 4\left(\frac{1}{\sqrt{2}} + i\frac{1}{\sqrt{2}}\right) = 2\sqrt{2} + 2\sqrt{2}i.$$

Answer: The rectangular form of the complex number is $2\sqrt{2} + 2\sqrt{2}i$.

Problem 2: Express $1+i\sqrt{3}$ in polar form.

You can express the complex number as $a+bi$ with $a=r\cos\theta$ and $b=r\sin\theta$. Let's use the formulas $r=\sqrt{x^2+y^2}$ and $\theta=\tan^{-1}\left(\dfrac{y}{x}\right)$ to solve for the r and θ values.

$$r=\sqrt{1^2+\left(\sqrt{3}\right)^2}=\sqrt{1+3}=\sqrt{4}=2$$

$$\theta=\tan^{-1}\left(\frac{y}{x}\right)=\tan^{-1}\left(\frac{\sqrt{3}}{1}\right)=\frac{\pi}{3}$$

Answer: The polar form of the complex number is $2cis\,\dfrac{\pi}{3}$.

Problem 3: Find the product of $3cis\,\dfrac{\pi}{3}$ and $4cis\,\dfrac{\pi}{6}$.

Recall that $z_1 z_2 = (r_1 cis\alpha)(r_2 cis\beta) = r_1 r_2 cis(\alpha+\beta)$. Thus, you need to multiply the absolute values of the two complex numbers and add their angles to get the answer.

$$3cis\,\frac{\pi}{3}\cdot4cis\,\frac{\pi}{6}=3\cdot4cis\left(\frac{\pi}{3}+\frac{\pi}{6}\right)=12cis\,\frac{\pi}{2}$$

Answer: The product of the two complex numbers is $12cis\,\dfrac{\pi}{2}$.

Problem 4: Show that $\left(-1+\sqrt{3}i\right)^{24}$ is a real number.

In order to show that the number is real, you must show that its imaginary component equals zero. Since you are dealing with a power of a complex number, De Moivre's Theorem comes in handy in providing an easy way to calculate the power of a complex number via the formula $z^n=r^n cisn\theta$.

However, we first need to convert the complex number into polar form. Since we are not interested in the absolute value of the complex number (since we only need to show that it's a real number), all we need to do is calculate θ.

$$\theta=\tan^{-1}\left(\frac{\sqrt{3}}{-1}\right)=\tan^{-1}\left(-\sqrt{3}\right)=-\frac{\pi}{3}$$

To show that the number is a real one, we need to show that its imaginary part $y=r\sin\theta$ is equal to zero. In our case, this would have to mean that $\sin 24\theta=0$.

$$\sin 24\theta=\sin(24\cdot-\frac{\pi}{3})=\sin(-8\pi)=0$$

Answer: Since we have shown that the imaginary part of the complex number is equal to zero, we have demonstrated that it is a real number.

Problem 5: Find the four fourth roots of –16.

Let's treat –16 as a complex number with an imaginary part of zero and express it in polar form. –16 becomes –16cis0. Using De Moivre's theorem, let's set up the problem as:

$$r^4 cis 4\alpha = -16 cis 0 = 16 \cdot - cis 0$$

By looking at the right side of the previous expression, you can conclude that:

$$r^4 = 16 \text{ or } r = 2$$

Also, $cis 4\alpha = -cis 0° = cis 180°$. This can be expressed as $cis(180° + \text{a multiple of } 360°)$ because the values of the sine and cosine will be the same for all these angles.

Therefore, $cis\alpha = cis 45°$ or (45° + a multiple of 90°). We are utilizing multiples of 90° since that measure is a fourth of 360°. Thus, $\alpha = 45°$, 45° + 90°, 45° + 180°, 45° + 270° or $\alpha = 45°$, 135°, 225°, and 315°. Because you need four fourth roots, you calculate only four angles.

Substitute the values for the angles' sines and cosines. Therefore, the four fourth roots of 8i are:

$$z_1 = 2cis 45° = 2(\cos 45° + i\sin 45°) = 2\left(\frac{1}{\sqrt{2}} + i\frac{1}{\sqrt{2}}\right) = \frac{2}{\sqrt{2}} + \frac{2}{\sqrt{2}}i$$

$$z_2 = 2cis 135° = 2(\cos 135° + i\sin 135°) = 2\left(-\frac{1}{\sqrt{2}} + i\frac{1}{\sqrt{2}}\right) = -\frac{2}{\sqrt{2}} + \frac{2}{\sqrt{2}}i$$

$$z_3 = 2cis 225° = 2(\cos 225° + i\sin 225°) = 2\left(-\frac{1}{\sqrt{2}} - i\frac{1}{\sqrt{2}}\right) = -\frac{2}{\sqrt{2}} - \frac{2}{\sqrt{2}}i$$

$$z_4 = 2cis 315° = 2(\cos 315° + i\sin 135°) = 2\left(\frac{1}{\sqrt{2}} - i\frac{1}{\sqrt{2}}\right) = \frac{2}{\sqrt{2}} - \frac{2}{\sqrt{2}}i$$

Answer: The fourth roots of –16 are $\frac{2}{\sqrt{2}} + \frac{2}{\sqrt{2}}i$, $-\frac{2}{\sqrt{2}} + \frac{2}{\sqrt{2}}i$, $-\frac{2}{\sqrt{2}} - \frac{2}{\sqrt{2}}i$, and $\frac{2}{\sqrt{2}} - \frac{2}{\sqrt{2}}i$.

Rules and Formulas Review

Geometry Definitions

- Complementary angles are two angles whose measures have the sum 90°.
- Supplementary angles are two angles whose measures have the sum 180°.

Triangle

- $A = \frac{1}{2}bh$, where A is the area, b is the base, and h is the height.
- The sum of all three angles equals 180°.
- Hero's Formula for the area:
  $A = \sqrt{s(s-a)(s-b)(s-c)}$, where a, b, and c are sides and $s = \frac{a+b+c}{2}$.

The Law of Sines

- In $\triangle ABC$, $\dfrac{\sin A}{a} = \dfrac{\sin B}{b} = \dfrac{\sin C}{c}$.

The Law of Cosines

- In $\triangle ABC$, $c^2 = a^2 + b^2 - 2ab\cos C$.

Circle

- Circumference $= C = 2\pi r$, where r is the radius.
- Area $= A = \pi r^2$, where r is the radius.

Sector of Circle

- When θ is in radians, then area = $\dfrac{\theta r^2}{2}$, $s = r\theta$.
- When θ is in degrees, then area = $\dfrac{\theta}{360} \cdot \pi r^2$, $s = \dfrac{\theta}{360} \cdot 2\pi r$.

Definition of the Six Trigonometric Functions

- Right triangle definitions where $0 < \theta < \dfrac{\pi}{2}$:

$$\sin\theta = \frac{\text{opp.}}{\text{hyp.}} \qquad \csc\theta = \frac{\text{hyp.}}{\text{opp.}}$$

$$\cos\theta = \frac{\text{adj.}}{\text{hyp.}} \qquad \sec\theta = \frac{\text{hyp.}}{\text{adj.}}$$

$$\tan\theta = \frac{\text{opp.}}{\text{adj.}} \qquad \cot\theta = \frac{\text{adj.}}{\text{opp.}}$$

- Circular function definitions, where θ is any angle and r is the radius:

$$\sin\theta = \frac{y}{r} \qquad \csc\theta = \frac{r}{y}$$

$$\cos\theta = \frac{x}{r} \qquad \sec\theta = \frac{r}{x}$$

$$\tan\theta = \frac{y}{x} \qquad \cot\theta = \frac{x}{y}$$

Reciprocal Identities

$$\sin\theta = \frac{1}{\csc\theta} \qquad \cos\theta = \frac{1}{\sec\theta} \qquad \tan\theta = \frac{1}{\cot\theta}$$

$$\cot\theta = \frac{1}{\tan\theta} \qquad \sec\theta = \frac{1}{\cos\theta} \qquad \csc\theta = \frac{1}{\sin\theta}$$

Tangent and Cotangent Identities

$$\tan\theta = \frac{\sin\theta}{\cos\theta} \qquad \cot\theta = \frac{\cos\theta}{\sin\theta}$$

Pythagorean Identities

$$\sin^2\theta + \cos^2\theta = 1$$

$$1 + \tan^2\theta = \sec^2\theta$$

$$1 + \cot^2\theta = \csc^2\theta$$

Cofunction Identities

$$\sin\theta = \cos\left(\frac{\pi}{2} - \theta\right) \qquad \cos\theta = \sin\left(\frac{\pi}{2} - \theta\right)$$

$$\tan\theta = \cot\left(\frac{\pi}{2} - \theta\right) \qquad \cot\theta = \tan\left(\frac{\pi}{2} - \theta\right)$$

$$\sec\theta = \csc\left(\frac{\pi}{2} - \theta\right) \qquad \csc\theta = \sec\left(\frac{\pi}{2} - \theta\right)$$

Negative Angle Identities

$$\sin(-\theta) = -\sin\theta \qquad \cos(-\theta) = \cos\theta$$

$$\tan(-\theta) = -\tan\theta \qquad \cot(-\theta) = -\cot\theta$$

$$\sec(-\theta) = \sec\theta \qquad \csc(-\theta) = -\csc\theta$$

Sum and Difference Formulas

$$\sin(u \pm v) = \sin u \cos v \pm \cos u \sin v$$

$$\cos(u \pm v) = \cos u \cos v \pm \sin u \sin v$$

$$\tan(u \pm v) = \frac{\tan u \pm \tan v}{1 \pm \tan u \tan v}$$

Double-Angle Formulas

$$\sin 2\theta = 2\sin\theta\cos\theta$$

$$\cos 2\theta = \cos^2\theta - \sin^2\theta = 2\cos^2\theta - 1 = 1 - 2\sin^2\theta$$

$$\tan 2\theta = \frac{2\tan\theta}{1 - \tan^2\theta}$$

Power–Reducing Formulas

$$\sin^2 \theta = \frac{1-\cos 2\theta}{2}$$

$$\cos^2 \theta = \frac{1+\cos 2\theta}{2}$$

$$\tan^2 \theta = \frac{1-\cos 2\theta}{1+\cos 2\theta}$$

Half-Angle Formulas

$$\sin \frac{u}{2} = \pm\sqrt{\frac{1-\cos u}{2}}$$

$$\cos \frac{u}{2} = \pm\sqrt{\frac{1+\cos u}{2}}$$

$$\tan \frac{u}{2} = \pm\sqrt{\frac{1-\cos u}{1+\cos u}} = \frac{\sin u}{1+\cos u} = \frac{1-\cos u}{\sin u}$$

Sum-to-Product Formulas

$$\sin u + \sin v = 2\sin\left(\frac{u+v}{2}\right)\cos\left(\frac{u-v}{2}\right)$$

$$\sin u - \sin v = 2\cos\left(\frac{u+v}{2}\right)\sin\left(\frac{u-v}{2}\right)$$

$$\cos u + \cos v = 2\cos\left(\frac{u+v}{2}\right)\cos\left(\frac{u-v}{2}\right)$$

$$\cos u - \cos v = -2\sin\left(\frac{u+v}{2}\right)\sin\left(\frac{u-v}{2}\right)$$

Product-to-Sum Formulas

$$\sin u \sin v = \frac{1}{2}\left[\cos(u-v) - \cos(u+v)\right]$$

$$\cos u \cos v = \frac{1}{2}\left[\cos(u-v) + \cos(u+v)\right]$$

$$\sin u \cos v = \frac{1}{2}\left[\sin(u+v) + \sin(u-v)\right]$$

$$\cos u \sin v = \frac{1}{2}\left[\sin(u+v) - \sin(u-v)\right]$$

Trigonometric Tables

The following tables can be utilized to calculate sine, cosine, tangent, cotangent, secant, and cosecant values of an angle given either in degrees or radians. Additionally, it can be utilized for conversion between degree values and numerical radian values.

For angles between 0° and 45°, you can look up the angle value in the left column and trace to the right in order to see corresponding values in the order depicted on the top row in the order of $\cos\theta$, $\sin\theta$, $\cot\theta$, $\tan\theta$, $\csc\theta$, and $\sec\theta$.

For angle values between 45° and 90°, you can look up the angle value in the right-most column and trace to the left in order to see trigonometric function values in the order of $\sec\theta$, $\csc\theta$, $\tan\theta$, $\cot\theta$, $\sin\theta$, and $\cos\theta$.

θ in Degrees	θ in Radians	$\sin\theta$	$\cos\theta$	$\tan\theta$	$\cot\theta$	$\sec\theta$	$\csc\theta$		
0°	0.0000	0.0000	1.0000	0.0000	Undef.	1.0000	Undef.	1.5708	90°
0.25°	0.0044	0.0044	1.0000	0.0044	229.18	1.0000	229.18	1.5664	89.75°
0.5°	0.0087	0.0087	1.0000	0.0087	114.59	1.0000	114.59	1.5621	89.5°
0.75°	0.0131	0.0131	0.9999	0.0131	76.390	1.0001	76.397	1.5577	89.25°
1°	0.0175	0.0175	0.9998	0.0175	57.290	1.0002	57.299	1.5533	89°
1.25°	0.0218	0.0218	0.9998	0.0218	45.829	1.0002	45.840	1.5490	88.75°
1.5°	0.0262	0.0262	0.9997	0.0262	38.188	1.0003	38.202	1.5446	88.5°
1.75°	0.0305	0.0305	0.9995	0.0306	32.730	1.0005	32.746	1.5403	88.25°
2°	0.0349	0.0349	0.9994	0.0349	28.636	1.0006	28.654	1.5359	88°
2.25°	0.0393	0.0393	0.9992	0.0393	25.452	1.0008	25.471	1.5315	87.75°
2.5°	0.0436	0.0436	0.9990	0.0437	22.904	1.0010	22.926	1.5272	87.5°
2.75°	0.0480	0.0480	0.9988	0.0480	20.819	1.0012	20.843	1.5228	87.25°
3°	0.0524	0.0523	0.9986	0.0524	19.081	1.0014	19.107	1.5184	87°
3.25°	0.0567	0.0567	0.9984	0.0568	17.611	1.0016	17.639	1.5141	86.75°
3.5°	0.0611	0.0610	0.9981	0.0612	16.350	1.0019	16.380	1.5097	86.5°
3.75°	0.0654	0.0654	0.9979	0.0655	15.257	1.0021	15.290	1.5053	86.25°
4°	0.0698	0.0698	0.9976	0.0699	14.301	1.0024	14.336	1.5010	86°
4.25°	0.0742	0.0741	0.9973	0.0743	13.457	1.0028	13.494	1.4966	85.75°
4.5°	0.0785	0.0785	0.9969	0.0787	12.706	1.0031	12.745	1.4923	85.5°
4.75°	0.0829	0.0828	0.9966	0.0831	12.035	1.0034	12.076	1.4879	85.25°
5°	0.0873	0.0872	0.9962	0.0875	11.430	1.0038	11.474	1.4835	85°
5.25°	0.0916	0.0915	0.9958	0.0919	10.883	1.0042	10.929	1.4792	84.75°
5.5°	0.0960	0.0958	0.9954	0.0963	10.385	1.0046	10.433	1.4748	84.5°
5.75°	0.1004	0.1002	0.9950	0.1007	9.9310	1.0051	9.9812	1.4704	84.25°

θ in Degrees	θ in Radians	$\sin \theta$	$\cos \theta$	$\tan \theta$	$\cot \theta$	$\sec \theta$	$\csc \theta$		
6°	**0.1047**	**0.1045**	**0.9945**	**0.1051**	**9.5144**	**1.0055**	**9.5668**	**1.4661**	**84°**
6.25°	0.1091	0.1089	0.9941	0.1095	9.1309	1.0060	9.1855	1.4617	83.75°
6.5°	0.1134	0.1132	0.9936	0.1139	8.7769	1.0065	8.8337	1.4573	83.5°
6.75°	0.1178	0.1175	0.9931	0.1184	8.4490	1.0070	8.5079	1.4530	83.25°
7°	**0.1222**	**0.1219**	**0.9925**	**0.1228**	**8.1443**	**1.0075**	**8.2055**	**1.4486**	**83°**
7.25°	0.1265	0.1262	0.9920	0.1272	7.8606	1.0081	7.9240	1.4443	82.75°
7.5°	0.1309	0.1305	0.9914	0.1317	7.5958	1.0086	7.6613	1.4399	82.5°
7.75°	0.1353	0.1349	0.9909	0.1361	7.3479	1.0092	7.4156	1.4355	82.25°
8°	**0.1396**	**0.1392**	**0.9903**	**0.1405**	**7.1154**	**1.0098**	**7.1853**	**1.4312**	**82°**
8.25°	0.1440	0.1435	0.9897	0.1450	6.8969	1.0105	6.9690	1.4268	81.75°
8.5°	0.1484	0.1478	0.9890	0.1495	6.6912	1.0111	6.7655	1.4224	81.5°
8.75°	0.1527	0.1521	0.9884	0.1539	6.4971	1.0118	6.5736	1.4181	81.25°
9°	**0.1571**	**0.1564**	**0.9877**	**0.1584**	**6.3138**	**1.0125**	**6.3925**	**1.4137**	**81°**
9.25°	0.1614	0.1607	0.9870	0.1629	6.1402	1.0132	6.2211	1.4094	80.75°
9.5°	0.1658	0.1650	0.9863	0.1673	5.9758	1.0139	6.0589	1.4050	80.5°
9.75°	0.1702	0.1693	0.9856	0.1718	5.8197	1.0147	5.9049	1.4006	80.25°
10°	**0.1745**	**0.1736**	**0.9848**	**0.1763**	**5.6713**	**1.0154**	**5.7588**	**1.3963**	**80°**
10.25°	0.1789	0.1779	0.9840	0.1808	5.5301	1.0162	5.6198	1.3919	79.75°
10.5°	0.1833	0.1822	0.9833	0.1853	5.3955	1.0170	5.4874	1.3875	79.5°
10.75°	0.1876	0.1865	0.9825	0.1899	5.2672	1.0179	5.3612	1.3832	79.25°
11°	**0.1920**	**0.1908**	**0.9816**	**0.1944**	**5.1446**	**1.0187**	**5.2408**	**1.3788**	**79°**
11.25°	0.1963	0.1951	0.9808	0.1989	5.0273	1.0196	5.1258	1.3744	78.75°
11.5°	0.2007	0.1994	0.9799	0.2035	4.9152	1.0205	5.0159	1.3701	78.5°
11.75°	0.2051	0.2036	0.9790	0.2080	4.8077	1.0214	4.9106	1.3657	78.25°

continues

θ in Degrees	θ in Radians	$\sin\theta$	$\cos\theta$	$\tan\theta$	$\cot\theta$	$\sec\theta$	$\csc\theta$		
12°	**0.2094**	**0.2079**	**0.9781**	**0.2126**	**4.7046**	**1.0223**	**4.8097**	**1.3614**	**78°**
12.25°	0.2138	0.2122	0.9772	0.2171	4.6057	1.0233	4.7130	1.3570	77.75°
12.5°	0.2182	0.2164	0.9763	0.2217	4.5107	1.0243	4.6202	1.3526	77.5°
12.75°	0.2225	0.2207	0.9753	0.2263	4.4194	1.0253	4.5311	1.3483	77.25°
13°	**0.2269**	**0.2250**	**0.9744**	**0.2309**	**4.3315**	**1.0263**	**4.4454**	**1.3439**	**77°**
13.25°	0.2313	0.2292	0.9734	0.2355	4.2468	1.0273	4.3630	1.3395	76.75°
13.5°	0.2356	0.2334	0.9724	0.2401	4.1653	1.0284	4.2837	1.3352	76.5°
13.75°	0.2400	0.2377	0.9713	0.2447	4.0867	1.0295	4.2072	1.3308	76.25°
14°	**0.2443**	**0.2419**	**0.9703**	**0.2493**	**4.0108**	**1.0306**	**4.1336**	**1.3265**	**76°**
14.25°	0.2487	0.2462	0.9692	0.2540	3.9375	1.0317	4.0625	1.3221	75.75°
14.5°	0.2531	0.2504	0.9681	0.2586	3.8667	1.0329	3.9939	1.3177	75.5°
14.75°	0.2574	0.2546	0.9670	0.2633	3.7983	1.0341	3.9277	1.3134	75.25°
15°	**0.2618**	**0.2588**	**0.9659**	**0.2679**	**3.7321**	**1.0353**	**3.8637**	**1.3090**	**75°**
15.25°	0.2662	0.2630	0.9648	0.2726	3.6680	1.0365	3.8018	1.3046	74.75°
15.5°	0.2705	0.2672	0.9636	0.2773	3.6059	1.0377	3.7420	1.3003	74.5°
15.75°	0.2749	0.2714	0.9625	0.2820	3.5457	1.0390	3.6840	1.2959	74.25°
16°	**0.2793**	**0.2756**	**0.9613**	**0.2867**	**3.4874**	**1.0403**	**3.6280**	**1.2915**	**74°**
16.25°	0.2836	0.2798	0.9600	0.2915	3.4308	1.0416	3.5736	1.2872	73.75°
16.5°	0.2880	0.2840	0.9588	0.2962	3.3759	1.0429	3.5209	1.2828	73.5°
16.75°	0.2923	0.2882	0.9576	0.3010	3.3226	1.0443	3.4699	1.2785	73.25°
17°	**0.2967**	**0.2924**	**0.9563**	**0.3057**	**3.2709**	**1.0457**	**3.4203**	**1.2741**	**73°**
17.25°	0.3011	0.2965	0.9550	0.3105	3.2205	1.0471	3.3722	1.2697	72.75°
17.5°	0.3054	0.3007	0.9537	0.3153	3.1716	1.0485	3.3255	1.2654	72.5°
17.75°	0.3098	0.3049	0.9524	0.3201	3.1240	1.0500	3.2801	1.2610	72.25°

θ in Degrees	θ in Radians	$\sin\theta$	$\cos\theta$	$\tan\theta$	$\cot\theta$	$\sec\theta$	$\csc\theta$		
18°	**0.3142**	**0.3090**	**0.9511**	**0.3249**	**3.0777**	**1.0515**	**3.2361**	**1.2566**	**72°**
18.25°	0.3185	0.3132	0.9497	0.3298	3.0326	1.0530	3.1932	1.2523	71.75°
18.5°	0.3229	0.3173	0.9483	0.3346	2.9887	1.0545	3.1515	1.2479	71.5°
18.75°	0.3272	0.3214	0.9469	0.3395	2.9459	1.0560	3.1110	1.2435	71.25°
19°	**0.3316**	**0.3256**	**0.9455**	**0.3443**	**2.9042**	**1.0576**	**3.0716**	**1.2392**	**71°**
19.25°	0.3360	0.3297	0.9441	0.3492	2.8636	1.0592	3.0331	1.2348	70.75°
19.5°	0.3403	0.3338	0.9426	0.3541	2.8239	1.0608	2.9957	1.2305	70.5°
19.75°	0.3447	0.3379	0.9412	0.3590	2.7852	1.0625	2.9593	1.2261	70.25°
20°	**0.3491**	**0.3420**	**0.9397**	**0.3640**	**2.7475**	**1.0642**	**2.9238**	**1.2217**	**70°**
20.25°	0.3534	0.3461	0.9382	0.3689	2.7106	1.0659	2.8892	1.2174	69.75°
20.5°	0.3578	0.3502	0.9367	0.3739	2.6746	1.0676	2.8555	1.2130	69.5°
20.75°	0.3622	0.3543	0.9351	0.3789	2.6395	1.0694	2.8225	1.2086	69.25°
21°	**0.3665**	**0.3584**	**0.9336**	**0.3839**	**2.6051**	**1.0711**	**2.7904**	**1.2043**	**69°**
21.25°	0.3709	0.3624	0.9320	0.3889	2.5715	1.0730	2.7591	1.1999	68.75°
21.5°	0.3752	0.3665	0.9304	0.3939	2.5386	1.0748	2.7285	1.1956	68.5°
21.75°	0.3796	0.3706	0.9288	0.3990	2.5065	1.0766	2.6986	1.1912	68.25°
22°	**0.3840**	**0.3746**	**0.9272**	**0.4040**	**2.4751**	**1.0785**	**2.6695**	**1.1868**	**68°**
22.25°	0.3883	0.3786	0.9255	0.4091	2.4443	1.0804	2.6410	1.1825	67.75°
22.5°	0.3927	0.3827	0.9239	0.4142	2.4142	1.0824	2.6131	1.1781	67.5°
22.75°	0.3971	0.3867	0.9222	0.4193	2.3847	1.0844	2.5859	1.1737	67.25°
23°	**0.4014**	**0.3907**	**0.9205**	**0.4245**	**2.3559**	**1.0864**	**2.5593**	**1.1694**	**67°**
23.25°	0.4058	0.3947	0.9188	0.4296	2.3276	1.0884	2.5333	1.1650	66.75°
23.5°	0.4102	0.3987	0.9171	0.4348	2.2998	1.0904	2.5078	1.1606	66.5°
23.75°	0.4145	0.4027	0.9153	0.4400	2.2727	1.0925	2.4830	1.1563	66.25°

continues

θ in Degrees	θ in Radians	$\sin\theta$	$\cos\theta$	$\tan\theta$	$\cot\theta$	$\sec\theta$	$\csc\theta$		
24°	**0.4189**	**0.4067**	**0.9135**	**0.4452**	**2.2460**	**1.0946**	**2.4586**	**1.1519**	**66°**
24.25°	0.4232	0.4107	0.9118	0.4505	2.2199	1.0968	2.4348	1.1476	65.75°
24.5°	0.4276	0.4147	0.9100	0.4557	2.1943	1.0989	2.4114	1.1432	65.5°
24.75°	0.4320	0.4187	0.9081	0.4610	2.1692	1.1011	2.3886	1.1388	65.25°
25°	**0.4363**	**0.4226**	**0.9063**	**0.4663**	**2.1445**	**1.1034**	**2.3662**	**1.1345**	**65°**
25.25°	0.4407	0.4266	0.9045	0.4716	2.1203	1.1056	2.3443	1.1301	64.75°
25.5°	0.4451	0.4305	0.9026	0.4770	2.0965	1.1079	2.3228	1.1257	64.5°
25.75°	0.4494	0.4344	0.9007	0.4823	2.0732	1.1102	2.3018	1.1214	64.25°
26°	**0.4538**	**0.4384**	**0.8988**	**0.4877**	**2.0503**	**1.1126**	**2.2812**	**1.1170**	**64°**
26.25°	0.4581	0.4423	0.8969	0.4931	2.0278	1.1150	2.2610	1.1126	63.75°
26.5°	0.4625	0.4462	0.8949	0.4986	2.0057	1.1174	2.2412	1.1083	63.5°
26.75°	0.4669	0.4501	0.8930	0.5040	1.9840	1.1198	2.2217	1.1039	63.25°
27°	**0.4712**	**0.4540**	**0.8910**	**0.5095**	**1.9626**	**1.1223**	**2.2027**	**1.0996**	**63°**
27.25°	0.4756	0.4579	0.8890	0.5150	1.9416	1.1248	2.1840	1.0952	62.75°
27.5°	0.4800	0.4617	0.8870	0.5206	1.9210	1.1274	2.1657	1.0908	62.5°
27.75°	0.4843	0.4656	0.8850	0.5261	1.9007	1.1300	2.1477	1.0865	62.25°
28°	**0.4887**	**0.4695**	**0.8829**	**0.5317**	**1.8807**	**1.1326**	**2.1301**	**1.0821**	**62°**
28.25°	0.4931	0.4733	0.8809	0.5373	1.8611	1.1352	2.1127	1.0777	61.75°
28.5°	0.4974	0.4772	0.8788	0.5430	1.8418	1.1379	2.0957	1.0734	61.5°
28.75°	0.5018	0.4810	0.8767	0.5486	1.8228	1.1406	2.0791	1.0690	61.25°
29°	**0.5061**	**0.4848**	**0.8746**	**0.5543**	**1.8040**	**1.1434**	**2.0627**	**1.0647**	**61°**
29.25°	0.5105	0.4886	0.8725	0.5600	1.7856	1.1461	2.0466	1.0603	60.75°
29.5°	0.5149	0.4924	0.8704	0.5658	1.7675	1.1490	2.0308	1.0559	60.5°
29.75°	0.5192	0.4962	0.8682	0.5715	1.7496	1.1518	2.0152	1.0516	60.25°

θ in Degrees	θ in Radians	sin θ	cos θ	tan θ	cot θ	sec θ	csc θ		
30°	**0.5236**	**0.5000**	**0.8660**	**0.5774**	**1.7321**	**1.1547**	**2.0000**	**1.0472**	**60°**
30.25°	0.5280	0.5038	0.8638	0.5832	1.7147	1.1576	1.9850	1.0428	59.75°
30.5°	0.5323	0.5075	0.8616	0.5890	1.6977	1.1606	1.9703	1.0385	59.5°
30.75°	0.5367	0.5113	0.8594	0.5949	1.6808	1.1636	1.9558	1.0341	59.25°
31°	**0.5411**	**0.5150**	**0.8572**	**0.6009**	**1.6643**	**1.1666**	**1.9416**	**1.0297**	**59°**
31.25°	0.5454	0.5188	0.8549	0.6068	1.6479	1.1697	1.9276	1.0254	58.75°
31.5°	0.5498	0.5225	0.8526	0.6128	1.6319	1.1728	1.9139	1.0210	58.5°
31.75°	0.5541	0.5262	0.8504	0.6188	1.6160	1.1760	1.9004	1.0167	58.25°
32°	**0.5585**	**0.5299**	**0.8480**	**0.6249**	**1.6003**	**1.1792**	**1.8871**	**1.0123**	**58°**
32.25°	0.5629	0.5336	0.8457	0.6310	1.5849	1.1824	1.8740	1.0079	57.75°
32.5°	0.5672	0.5373	0.8434	0.6371	1.5697	1.1857	1.8612	1.0036	57.5°
32.75°	0.5716	0.5410	0.8410	0.6432	1.5547	1.1890	1.8485	0.9992	57.25°
33°	**0.5760**	**0.5446**	**0.8387**	**0.6494**	**1.5399**	**1.1924**	**1.8361**	**0.9948**	**57°**
33.25°	0.5803	0.5483	0.8363	0.6556	1.5253	1.1958	1.8238	0.9905	56.75°
33.5°	0.5847	0.5519	0.8339	0.6619	1.5108	1.1992	1.8118	0.9861	56.5°
33.75°	0.5890	0.5556	0.8315	0.6682	1.4966	1.2027	1.8000	0.9817	56.25°
34°	**0.5934**	**0.5592**	**0.8290**	**0.6745**	**1.4826**	**1.2062**	**1.7883**	**0.9774**	**56°**
34.25°	0.5978	0.5628	0.8266	0.6809	1.4687	1.2098	1.7768	0.9730	55.75°
34.5°	0.6021	0.5664	0.8241	0.6873	1.4550	1.2134	1.7655	0.9687	55.5°
34.75°	0.6065	0.5700	0.8216	0.6937	1.4415	1.2171	1.7544	0.9643	55.25°
35°	**0.6109**	**0.5736**	**0.8192**	**0.7002**	**1.4281**	**1.2208**	**1.7434**	**0.9599**	**55°**
35.25°	0.6152	0.5771	0.8166	0.7067	1.4150	1.2245	1.7327	0.9556	54.75°
35.5°	0.6196	0.5807	0.8141	0.7133	1.4019	1.2283	1.7221	0.9512	54.5°
35.75°	0.6240	0.5842	0.8116	0.7199	1.3891	1.2322	1.7116	0.9468	54.25°

continues

θ in Degrees	θ in Radians	$\sin\theta$	$\cos\theta$	$\tan\theta$	$\cot\theta$	$\sec\theta$	$\csc\theta$		
36°	**0.6283**	**0.5878**	**0.8090**	**0.7265**	**1.3764**	**1.2361**	**1.7013**	**0.9425**	**54°**
36.25°	0.6327	0.5913	0.8064	0.7332	1.3638	1.2400	1.6912	0.9381	53.75°
36.5°	0.6370	0.5948	0.8039	0.7400	1.3514	1.2440	1.6812	0.9338	53.5°
36.75°	0.6414	0.5983	0.8013	0.7467	1.3392	1.2480	1.6713	0.9294	53.25°
37°	**0.6458**	**0.6018**	**0.7986**	**0.7536**	**1.3270**	**1.2521**	**1.6616**	**0.9250**	**53°**
37.25°	0.6501	0.6053	0.7960	0.7604	1.3151	1.2563	1.6521	0.9207	52.75°
37.5°	0.6545	0.6088	0.7934	0.7673	1.3032	1.2605	1.6427	0.9163	52.5°
37.75°	0.6589	0.6122	0.7907	0.7743	1.2915	1.2647	1.6334	0.9119	52.25°
38°	**0.6632**	**0.6157**	**0.7880**	**0.7813**	**1.2799**	**1.2690**	**1.6243**	**0.9076**	**52°**
38.25°	0.6676	0.6191	0.7853	0.7883	1.2685	1.2734	1.6153	0.9032	51.75°
38.5°	0.6720	0.6225	0.7826	0.7954	1.2572	1.2778	1.6064	0.8988	51.5°
38.75°	0.6763	0.6259	0.7799	0.8026	1.2460	1.2822	1.5976	0.8945	51.25°
39°	**0.6807**	**0.6293**	**0.7771**	**0.8098**	**1.2349**	**1.2868**	**1.5890**	**0.8901**	**51°**
39.25°	0.6850	0.6327	0.7744	0.8170	1.2239	1.2913	1.5805	0.8858	50.75°
39.5°	0.6894	0.6361	0.7716	0.8243	1.2131	1.2960	1.5721	0.8814	50.5°
39.75°	0.6938	0.6394	0.7688	0.8317	1.2024	1.3007	1.5639	0.8770	50.25°
40°	**0.6981**	**0.6428**	**0.7660**	**0.8391**	**1.1918**	**1.3054**	**1.5557**	**0.8727**	**50°**
40.25°	0.7025	0.6461	0.7632	0.8466	1.1812	1.3102	1.5477	0.8683	49.75°
40.5°	0.7069	0.6494	0.7604	0.8541	1.1708	1.3151	1.5398	0.8639	49.5°
40.75°	0.7112	0.6528	0.7576	0.8617	1.1606	1.3200	1.5320	0.8596	49.25°
41°	**0.7156**	**0.6561**	**0.7547**	**0.8693**	**1.1504**	**1.3250**	**1.5243**	**0.8552**	**49°**
41.25°	0.7199	0.6593	0.7518	0.8770	1.1403	1.3301	1.5167	0.8508	48.75°
41.5°	0.7243	0.6626	0.7490	0.8847	1.1303	1.3352	1.5092	0.8465	48.5°
41.75°	0.7287	0.6659	0.7461	0.8925	1.1204	1.3404	1.5018	0.8421	48.25°

θ in Degrees	θ in Radians	sin θ	cos θ	tan θ	cot θ	sec θ	csc θ		
42°	**0.7330**	**0.6691**	**0.7431**	**0.9004**	**1.1106**	**1.3456**	**1.4945**	**0.8378**	**48°**
42.25°	0.7374	0.6724	0.7402	0.9083	1.1009	1.3510	1.4873	0.8334	47.75°
42.5°	0.7418	0.6756	0.7373	0.9163	1.0913	1.3563	1.4802	0.8290	47.5°
42.75°	0.7461	0.6788	0.7343	0.9244	1.0818	1.3618	1.4732	0.8247	47.25°
43°	**0.7505**	**0.6820**	**0.7314**	**0.9325**	**1.0724**	**1.3673**	**1.4663**	**0.8203**	**47°**
43.25°	0.7549	0.6852	0.7284	0.9407	1.0630	1.3729	1.4595	0.8159	46.75°
43.5°	0.7592	0.6884	0.7254	0.9490	1.0538	1.3786	1.4527	0.8116	46.5°
43.75°	0.7636	0.6915	0.7224	0.9573	1.0446	1.3843	1.4461	0.8072	46.25°
44°	**0.7679**	**0.6947**	**0.7193**	**0.9657**	**1.0355**	**1.3902**	**1.4396**	**0.8029**	**46°**
44.25°	0.7723	0.6978	0.7163	0.9742	1.0265	1.3961	1.4331	0.7985	45.75°
44.5°	0.7767	0.7009	0.7133	0.9827	1.0176	1.4020	1.4267	0.7941	45.5°
44.75°	0.7810	0.7040	0.7102	0.9913	1.0088	1.4081	1.4204	0.7898	45.25°
45°	**0.7854**	**0.7071**	**0.7071**	**1.0000**	**1.0000**	**1.4142**	**1.4142**	**0.7854**	**45°**
		cos θ	sin θ	cot θ	tan θ	csc θ	sec θ	θ in Radians	θ in Degrees

Index

B-C

T